Propositions

THE RUTGERS LECTURES IN PHILOSOPHY

Larry Temkin, series editor

RUTGERS
School of Arts and Sciences

Published in the series

Vagueness: A Global Approach
Kit Fine

Freedom of Speech and Expression:
Its History, Its Value, Its Good Use, and Its Misuse
Richard Sorabji

Propositions: Ontology and Logic
Robert Stalnaker

Propositions

Ontology and Logic

ROBERT STALNAKER

OXFORD
UNIVERSITY PRESS

Oxford University Press is a department of the University of Oxford. It furthers the University's objective of excellence in research, scholarship, and education by publishing worldwide. Oxford is a registered trade mark of Oxford University Press in the UK and certain other countries.

Published in the United States of America by Oxford University Press
198 Madison Avenue, New York, NY 10016, United States of America.

Library of Congress Cataloging-in-Publication Data
Names: Stalnaker, Robert, author.
Title: Propositions : ontology and logic / Robert Stalnaker.
Description: New York, NY : Oxford University Press, [2023] |
Includes bibliographical references and index.
Identifiers: LCCN 2021062851 (print) | LCCN 2021062852 (ebook) |
ISBN 9780197647035 (hardback) | ISBN 9780197647059 (epub)
Subjects: LCSH: Proposition (Logic)
Classification: LCC BC181 .S68 2022 (print) | LCC BC181 (ebook) |
DDC 160—dc23/eng/20220315
LC record available at https://lccn.loc.gov/2021062851
LC ebook record available at https://lccn.loc.gov/2021062852

DOI: 10.1093/oso/9780197647035.001.0001

Printed by Integrated Books International, United States of America

Contents

Series Editor Forward

In 2014, I had the distinct privilege of being Chair of the Philosophy Department at Rutgers, The State University of New Jersey. Soon after becoming Chair, Peter Ohlin, the Philosophy Editor of Oxford University Press America, broached the idea of Rutgers organizing a major annual lecture series, where Rutgers would carefully select among the world's leading philosophers to give a series of three original lectures which would be subsequently revised for publication in a special book series by Oxford University Press. It was Peter's hope that such a Series might soon come to be recognized as one of the most important annual events on the philosophical calendar, and that it might one day come to rival the Locke Lectures, Dewey Lectures, and Tanner Lectures in stature.

I shared Peter's enthusiasm for the idea and promised to take it up with my colleagues and our administration. Unsurprisingly, the response was overwhelmingly supportive. However, it was decided that if we were going to take on such a commitment, we wanted the endeavor to involve much more than just adding another three lectures, each year, to our already crowded academic calendar. We wanted to create a Series that was truly distinctive and special; one that would benefit not only our faculty, but our outstanding graduate students, undergraduates, and the wider Rutgers community. Moreover, we were especially concerned to create a Series that would be both personally and intellectually rewarding for our visiting speakers.

After many hours of discussion with the Department and the Administration, the contours of the *Rutgers Lectures in Philosophy Series* took shape. First, the Department was committed to bringing in genuinely world-class philosophers who had already done, and were continuing to do, seminal work that was profoundly impacting their areas of research and, in some cases, the world at large. Our hope, and expectation, was that the lectures they would be delivering at Rutgers, and the books based on those lectures, would help set the philosophical agendas in their respective fields for many years to come. Bearing that goal foremost in mind, we were also committed to inviting a diverse group of speakers representing a broad spectrum of philosophical areas and interests within the analytic tradition.

Second, speakers would be asked to be available on campus for a full week and to pitch their first lecture so that interested alumni, the broader Rutgers academic community, and the general public might benefit from these internationally renowned philosophers. Their two subsequent lectures would be aimed at the high philosophical level that would most benefit the philosophy faculty, graduate students, visitors, and speaker, and would likely form the backbone of the speaker's subsequent book. Each lecture would be followed by discussion, a reception, and dinner, to facilitate serious engagement with the speaker's views.

Third, speakers would distribute drafts of their lectures, which would be read and discussed in advance by interested faculty, graduate students, and visitors. Then, there would be an intensive workshop with the speaker on the distributed material during their week on campus, providing the speaker with valuable feedback, and the workshop members with the chance to get to know the speaker and his or her views in a seminar setting.

Fourth, there would be a separate lunch meeting with the speaker for interested undergraduate majors, minors, philosophy club members, and students working on the undergraduate philosophy journal, *Arête*. As it turns out, several speakers have reported their meeting with our undergraduates to be one of the biggest highlights of their visits!

Finally, speakers would be encouraged to hang around the Department for impromptu discussion, and to set up individual meetings with faculty or graduate students particularly interested in their work.

A little over two years after Peter Ohlin and I first spoke, the first Rutgers Lectures in Philosophy Series took place, and at this point the first five lectures in the Series have already been delivered. And now this third book in the Series joins the first two that are already out, Kit Fine's *Vagueness: A Global Approach* and Sir Richard Sorabji's *Freedom of Speech and Expression: Its History, Its Value, Its Good Use, and Its Misuse*. It has been a long road between Peter Ohlin's original idea and this moment, but we couldn't be more pleased by how the Series has developed. It has truly been everything we envisaged it to be and more.

There are many people I would like to thank for helping to make the Rutgers Lectures in Philosophy Series possible. I apologize, in advance, for the fact that nothing I say here can remotely reflect the depth of my gratitude to those who have worked tirelessly behind the scenes to make this Series a success. I also apologize to anyone I inadvertently forget to mention. On both scores, please forgive me.

First, my deep gratitude to Peter Ohlin, who not only planted the seed from which this Series has grown, but who has helped to nurture it and bring it to fruition at every step of the way. Peter's contributions to this Series have been both invaluable and indispensable. My thanks, also, to Peter's excellent team at Oxford University Press, from members of the Editorial Board, to copyeditors, design artists, publicists, production managers, and so on, all of whom do yeomen's work in bringing a book to press and making sure that it receives the attention it deserves. Thanks, also, to Peter Momtchiloff, the head of Oxford University Press's Philosophy Division in the United Kingdom, for sage advice in this project's initial stages. And a special thanks to the referees for the volumes in this Series, whose conscientious comments are almost always underappreciated by everyone but the authors.

In the Philosophy Department, my biggest debt is to my colleagues, who were extremely supportive of this project from the get-go, and who have spent countless hours deliberating about the Series format and possible speakers, and attending the numerous Series events. I also want to thank all the graduate students who have actively participated in Series events, with a special thanks to Jimmy Goodrich and Adam Gibbons, who have done much of the heavy lifting in organizing the Series. I also owe a debt of gratitude to two business managers, Pauline Mitchell and Charlene Jones, our undergraduate administrative assistants, Jean Urteil and Jessica Koza, and, especially, our graduate administrative assistant, Mercedes Diaz, for all their hard work on behalf of this Series. Justin Kalef has organized the undergraduate luncheon meetings, and I am grateful to him and all the undergraduates who have made that event one of the Series highlights. Special thanks is also owed to Dean Zimmerman and Karen Bennett, who followed me as Chair of the Rutgers Philosophy Department, and both of whom, in turn, have offered unwavering support of the Series.

This Series would never exist were it not for the substantial support that the Philosophy Department has received over many years from the Rutgers Administration, including presidents, chancellors, vice-presidents, and deans. The Rutgers Lectures in Philosophy Series thrives because of the incredibly vibrant and congenial philosophical environment that the Rutgers Administration has made possible. I, and my colleagues, can't thank them enough. I can't possibly name every important administrator whose support has helped make the Rutgers Philosophy Department what it is today, but I need to mention a few. First, and foremost, the former president of the university, Robert Barchi, was committed to excellence and was a great champion of our department. Thankfully, our current president, Jonathan Holloway, is also committed to the success of our department.

Several people in the Dean's office have worked very hard to give the Rutgers Lectures in Philosophy Series a high profile not only within Rutgers, but globally. These include Kara Donaldson, who oversaw these efforts, Ian DeFalco, and John Chadwick. Michelle Stephens was supportive of the Series as Dean of Humanities, and that support continues with our current Dean of Humanities, Rebecca Walkowitz. Two people are deserving of special mention, James Swenson, former Dean of Humanities, and currently Vice Provost for Academic Affairs, and Peter March, Executive Dean of the School of Arts and Sciences. To each, my heartfelt thanks for all they have done over many years to support the Philosophy Department in general, and the Lecture Series in particular.

Finally, my biggest thanks go to the stars of the Series, the speakers whose path-breaking work is an inspiration to all those in the field. I am most grateful to those who helped put the Series on the philosophical map, by agreeing to be one of our distinguished speakers. As of now, the lineup of committed speakers for the Series is as follows: Kit Fine (2016), Sir Richard Sorabji (2017), Robert Stalnaker (2018), Jeff McMahan (2019), Béatrice Longuenesse (2020), Tim Williamson (2021), Philip Kitcher (2022), and Susanna Siegel (2023).

Robert Stalnaker has been a delight to work with from the first moment I asked him to participate in our series through delivery of the final page proofs. The three lectures he delivered at Rutgers, "Propositions," "Predicates and Predication," and "Properties and Relations" were everything we were hoping for from a speaker in our series. Wielding an extremely insightful and rigorous set of arguments, Stalnaker offered a powerful approach to thinking about a set of fundamentally important issues at the intersection of philosophy of language, logic, and metaphysics. Moreover, throughout his visit, Stalnaker was a model of generosity and congeniality, whether he was engaged with faculty, graduate students, or undergraduates. The book you are now reading, *Propositions: Ontology and Logic*, is certain to be much discussed for many years to come. I am extremely pleased to have Robert Stalnaker's book appear as the third book in the Rutgers Lectures in Philosophy Series. Stalnaker's book is another important step toward the realization of our goal, that one day our Series would be one of the most important in all of Philosophy.

Again, to everyone who has helped in bringing this book to press, and most especially Robert Stalnaker, my heartfelt gratitude.

Larry S. Temkin
Series Editor
New Brunswick, New Jersey
September 2021

Acknowledgments

This book grew out of lectures delivered at Rutgers in October 2018 as part of an annual series, the Rutgers lectures in philosophy. I am grateful to the Rutgers department, and particularly to Larry Temkin, for the invitation that gave me the opportunity to develop these ideas. Thanks also to the audiences for those lectures for their comments and questions.

I have always received valuable advice about my work from students and colleagues, but in the case of these lectures and the manuscript that developed from them, readers have been particularly generous and insightful. I received detailed comments from a number of people on the typescripts of the lectures and on several drafts of the chapters of the book, comments that led to significant revisions and corrections, both of the general development of the philosophical picture and of the technical details of many of the arguments. I want particularly to thank five people whose comments, written and in discussion, had a major effect on my understanding of the issues and on the shape of the book.

Even before I began on this current project, Peter Fritz had already had a big impact on my thinking about the issues discussed in the lectures through his comments on some technical arguments in an appendix to an earlier book of mine,[1] comments that pointed out and corrected some mistakes in those arguments, and showed how to give them a much clearer formulation. This book aims to develop further the ideas in that earlier book, and Fritz commented with great care and detail on several drafts of it. He advised me to spell

[1] Stalnaker 2012. Fritz's comments were published in Fritz 2016.

out the details of the proofs of the formal claims I made, including a completeness theorem for the modal logic, and proof sketches of object-language theorems used in that proof. I took his advice, which proved well taken, since interesting subtleties emerged from the details. His meticulous comments helped me to avoid some mistakes and to get a better understanding of both technical and philosophical issues.

Jeremy Goodman read the penultimate draft of the whole manuscript and gave me an extensive set of comments on all aspects of the book, comments that showed the philosophical subtlety and ingenuity that we have come to expect from him. I made significant changes in response to his challenges. I describe my defense of propositions, and of modal logic, as a Quinean defense, despite the fact that it defends a number of theses that Quine was famous for rejecting. Goodman questioned the Quinean framing, and he urged me either to downplay it or to develop and defend it in more detail. I took up this challenge, taking the latter course, aiming to clarify just what features of the Quinean story I meant to be drawing on and to spell out in more detail the overall dialectic of my argument. Goodman's comments on substantive issues and arguments showed me that there is a different way of looking at the metaphysical landscape in which a theory of propositions, properties, and relations is developed, and that the differences run deep. I do not yet have a full understanding of all aspects of the alternative picture, some of which are shared with each of the other interlocutors that I am mentioning here. I hope the development of my own approach will help to bring the issues into sharper focus, but direct comparison of the different approaches will be a task for further work. In any case, Goodman's stimulating comments deepened my understanding of the issues, and I hope they led to improvements in the book.

Nick Jones, who was visiting MIT during the time I was revising the manuscript, gave me incisive comments on the typescripts of the lectures, and we had a number of good conversations about the issues. The focus of our disagreements concerned the interpretation of higher-order modal logic—type theory—and its application to metaphysical issues. I argue in the book that this kind of logical framework, interpreted a certain way, can be reconciled with a Quinean methodology, but Jones argued for a different interpretation of that framework that diverges more radically from the Quinean picture. I was not convinced, but I did get a clearer view of what the alternative interpretation is. And I was convinced (as Quine was) that a first-order theory with an ontology of properties and relations of all the different types faces a threat of paradox. The discussion of this in Chapter 5 is a response to Jones's argument.

Bruno Jacinto read the whole manuscript and gave me excellent comments that led to a number of revisions and clarifications. Jacinto also influenced the book through his published papers on contingentist metaphysics, higher-order logics, and the interpretation of names. We had an extensive email discussion in which he very patiently helped me to understand a tension between contingentist metaphysics and a straightforward compositional semantics for quantified modal logic.

I have been discussing a wide range of philosophical issues with Agustin Rayo since he was a graduate student at MIT, and especially after he joined the faculty here. The project carried out in this book was influenced by those discussions and by my reading of his book, Rayo 2013, *The Construction of Logical Space,* which develops a pluralistic account of ontology that I find congenial. He read more than one draft of the manuscript of this book and gave me very helpful comments at several stages. In particular, he helped me to see the points at which I was making choices about how to represent predication and properties that had consequences for the metaphysics.

Thanks to Peter Ohlin, my editor at Oxford University Press, for his support and advice. I enjoyed working with him for some years when I was one of the OUP philosophy delegates and was glad to have the opportunity to work with him on a book. Thanks, more generally to the Press for their support of the Rutgers Lectures in Philosophy series.

Finally, for editorial help at the late stages, including the creation of the index, thanks to Joshua Pearson.

Introduction

This book is about propositions, properties, and relations, and how to theorize about them. It is part of a defense of the metaphysical thesis that has come to be called "contingentism," or more precisely, a defense of a theoretical framework that allows for the coherent development of that thesis. Contingentism, informally characterized, consists of a pair of theses: first, that some things might have failed to exist, including, perhaps, abstract objects such as propositions, properties, and relations; and second, that there might have been things of all these kinds other than those that actually do exist. Timothy Williamson, who coined the term, gave it a formal definition in modal logic as follows: Contingentism is the negation of *necessitism*, which is the thesis, $\Box\forall x\Box\exists y\, y=x$.[1] The motivation for contingentism, I will argue, has its roots in a Quinean naturalistic philosophical picture, and my project in this book will take the form of a conversation with Quine, drawing on his work and his ideas, but also arguing that he should find more congenial than he did the conceptual resources that are necessary for making the thesis of contingentism intelligible.

Quine began his book on the philosophy of logic with a series of objections to the hypothesis that there are such things as propositions, and he moves quickly to a section called "propositions dismissed." Properties and relations (what Quine called "attributes") are also decisively rejected for similar reasons. Quine also held

[1] Williamson 2013. Williamson defends necessitism. Contingentism (though not by that name) is defended by Robert Adams and Kit Fine, and by me. See Fine 1977 and 2002, Adams 1981, and Stalnaker 2012.

Propositions. Robert Stalnaker, Oxford University Press. © Oxford University Press 2023.
DOI: 10.1093/oso/9780197647035.003.0001

that logic has no room for a *modal* logic, a logic with operators expressing necessity and possibility. He argued that the development of modal logic was based on a use-mention confusion. But despite the fact that the theses and concepts I will defend were emphatically rejected by Quine, I am describing my project as an attempt to give a Quinean foundation for modal logic, and for a theory of propositions, properties, and relations. Even if the project starts with a stark contrast with Quine in ontological commitment, and even though it winds up using conceptual resources (modal and higher-order logics) that Quine rejected, I will adopt a Quinean methodology, following him in giving primacy to first-order extensional logic, and throughout keeping my eye on distinctions that he emphasized, and trying to avoid the confusions that he warned against. At one point in the *Philosophy of Logic*, Quine contrasts the confused logician with the "unconfused but prodigal logician who embraces attributes with his eyes open."[2] I will do my best to follow the latter's example.

To set the stage, I will start in Chapter 1 by sketching in broad strokes Quine's philosophical picture, as I understand it. I will look at his critique of some of the basic concepts of the logical empiricist tradition, and at the consequences, both substantive and methodological, that he took that critique to have. I will sketch his well-articulated conception of the kind of framework in which philosophical analysis should be done, and his ideas about the scope and nature of logic.

In Chapter 2, I will describe a conception of *proposition* and look at Quine's reasons for dismissing propositions from his ontology. I will argue that we can meet his objections and can reconcile the account of propositions with Quine's general requirements for an acceptable ontological commitment, though I will acknowledge that the conception of proposition I will defend does not do all the philosophical work that propositions have been asked to do.

[2] Quine 1986, 66.

After describing and defending a simple theory of propositions, I will make the seemingly innocent assumption that if there are propositions, then the sentences of the language of our theory of propositions themselves *express* propositions. More generally, if there are propositions, we should assume that the sentences of the kind of regimented language that a Quinean project aims to construct express propositions. This assumption will complicate our theory, requiring an enrichment of the expressive resources of the first-order language in which the theory was formulated, and raising a number of new questions. But the assumption allows us, without further commitment, to introduce modal operators to the language of the theory and to provide them with a semantics. So I will digress at that point to look at Quine's criticisms of modal logic, arguing that while they are justified criticisms of the way modal logic was interpreted by its early developers, they do not apply in the context of the theory of propositions that I am proposing. The enrichment of the ontology and the assumption that sentences of the language *express* as well as *describe* propositions, however, do raise questions about the logic of propositions that the principles of the basic theory do not answer. These further questions focus attention on the *predicates* of the theory of propositions, and more generally on conditions for the admissibility for predicates to the kind of regimented language that Quine wanted to use to formulate his views about ontology. Questions about predicates and predication will be the topic of Chapter 3.

Even though admitting propositions to our ontology raises new questions about predicates and predication, I will argue in the third chapter that these are questions that a Quinean needs to answer in any case, and that the resources of the theory of propositions help to sharpen them by clarifying the commitments that are involved in the judgment that a predicate is admissible to a regimented language. Just as (according to Quine's naturalistic picture) the construction of a regimented language for describing the world requires substantive presuppositions about ontology (about what there is to talk about) so it requires substantive presuppositions about ideology (about what resources are available for describing the thing in one's ontology). The role of predicates in determining the propositions expressed with them helps to provide criteria for assessing their admissibility and to bring out the presuppositions of admitting them.

After discussing the general problems about what needs to be said about predicates and names to justify admitting them to an acceptable formalized language, I will conclude Chapter 3 by looking at the specific predicates of the theory of propositions and at the propositions about propositions that are expressed with those predicates.

Chapter 4 is a technical digression, first spelling out the details of the first-order modal logic that I am using, and of its model theory, and then defining a specific model of this kind for the theory of propositions. One aim of this chapter is to put the details of the model theory on the table in order to set up a later discussion of the relation between the Kripke models for the theory and the reality they purport to model. A second aim is to show that we can have a consistent theory that *expresses* propositions and is also *about* a domain of propositions that includes those expressed in that theory. Specifically, we show that there is no paradox in a theory that has a truth predicate that applies to the propositions expressible in it.

One upshot of the general discussion in Chapter 3 of the predicates of a regimented language is that we can specify something like identity conditions for the roles played by predicates in the determination of propositions. Just as an ontological commitment presupposes that there are answers to questions about when two referring expressions refer to the same thing, so the assumption that the sentences of a first-order language expresses propositions presupposes that there are answers to questions about when two predicates have the same compositional role in determining propositions. It is tempting to think that when presuppositions of this kind are satisfied, that is enough to justify an ontological commitment to entities that correspond to predicates—properties and relations—but the discussion of predication in Chapter 3 stops short of endorsing that commitment. In Chapter 5, I explore the consequences of taking this further step, and I consider the character of a regimented language that is appropriate for stating a theory about the relationships among individuals, propositions, properties, and relations, if such there be. The hypothesis I will explore is that when a predicate is admissible to a regimented language, there exists a property or relation of the appropriate kind that is expressed by that predicate. This move will be a further commitment, and as we will see, it will require some qualification and some care in defining and interpreting a formal language that is appropriate for this purpose. The aim will be to clarify just what kinds of things we are talking about when we talk about properties and relations, how they are related to each other, and what kind of language it is best to use to state a theory of properties and relations. We will follow, in Chapter 5, the standard practice of using a higher-order language to formulate the theory. Quine warned that a language of this kind blurred some important distinctions and made it easier to equivocate. I will take his warning seriously but argue that we can, with care, avoid the equivocations. I will also consider, in the last part of the chapter, the possibility of formulating the theory as a first-order theory with the whole hierarchy of individuals, propositions, properties, and relations as its domain. This kind of formulation presents a threat of paradox that is more serious than the threat of paradox for a first-order theory of propositions that is discussed in Chapter 4, but I will argue that the paradox can be defused, and that this alternative way of regimenting a theory of properties and relations brings to the surface some questions that need to be answered.

Chapter 6 is about the relation between the Kripke model theory used to give the compositional semantics for the modal logic and the reality (the individuals, propositions, properties, and relations) that the theories being interpreted are about. The ontological commitments we are endorsing are more extravagant than Quine would have liked, but they do not include a commitment to merely possible worlds, or merely possible individuals. This chapter aims to explain how we can reconcile the use of models of this kind to do the compositional semantics for our regimented modal language with the rejection of some of the ontological commitment that the model theory seems to be presupposing.

1
The Quinean Legacy

The ideas I will develop and defend in later chapters are neo-Quinean, but they will diverge in significant ways from views that Quine held. In this first chapter, the exposition will stay closer to Quine's actual views, though my interpretation of Quine will be opinionated and selective. My aim is to lay out the common ground against which I will later argue for a reorientation of some aspects of the Quinean picture.[1]

1.1. Frameworks and regimentation

Rudolf Carnap was Quine's mentor, and his most important interlocutor in the development of his own philosophical position. Quine was sympathetic with Carnap's pragmatism, and with the idea that a formal language based on extensional first-order quantification theory was the appropriate framework for formulating one's conception of what there is and for addressing problems of philosophical analysis. But because he rejected the analytic/synthetic distinction that was central to Carnap's projects, he thought that the foundations of this method of philosophical analysis needed to be rethought.

In his influential paper, "Empiricism, Semantics and Ontology," Carnap had made a sharp distinction between the *decision* to accept

[1] My interpretation is informed and influenced by a recent excellent scholarly study of the development of Quine's naturalism by Sander Verhaegh. See Verhaegh 2018.

Propositions. Robert Stalnaker, Oxford University Press. © Oxford University Press 2023.
DOI: 10.1093/oso/9780197647035.003.0002

a certain framework (either empirical or mathematical) for doing philosophy and the *judgments* that one makes within the framework (once it has been accepted) about what the application of the rules that come with the framework tell us, more specifically, about what there is and about what is true. The acceptance of a framework was, for Carnap, the acceptance of a certain ontology, but he thought it was a mistake to think of the decision to accept an ontology as a theoretical judgment about what there is in the world to talk about.

> If someone decides to accept the thing language, there is no objection against saying that he has accepted the world of things. But this must not be interpreted as if it meant his acceptance of a *belief* in the reality of the thing world: there is no such belief or assertion or assumption, because it is not a theoretical question. To accept the thing world means nothing more than to accept a certain form of language.[2]

Once one has adopted the thing language as one's basic empirical framework, one can then consider questions about which statements expressed in that language are true (including questions about what scientific theories are correct), but those questions presuppose the basic framework, with its ontology, and make no sense outside of it.

Quine agreed with Carnap that the decision to accept a theoretical framework was based on a practical judgment that doing so was a fruitful way of organizing and predicting one's experience and finding one's way around one's environment, but he rejected the distinction between external practical questions about what language to adopt and internal questions about what theory to accept as correct. "Carnap maintains that ontological questions,

[2] Carnap 1950.

and likewise questions of logical or mathematical principle, are questions, not of fact, but of choosing a convenient scheme or framework for science; and with this I agree only if the same be conceded for every scientific hypothesis."[3] This concluding remark from Quine's critique of Carnap's views on ontology might seem to suggest that he was accepting Carnap's anti-realism about frameworks, extending it to a pragmatic anti-realism about theoretical judgments generally, but the dialectical strategy of Quine's critique of Carnap was the opposite of this: his aim was to defend *realism* about basic ontological judgments by questioning the line between decisions about what linguistic framework to work within, and decisions about what to accept as true. The strategy is developed most explicitly in a paper, "Posits and Reality," written in about 1955[4] There he began with the familiar point that scientific judgments, such as that material bodies are made up of molecules, are supported by evidence only indirectly: they are hypotheses that are accepted because they help to organize our observations of ordinary things and to predict their behavior. The temptation, Quine suggested, is to take a theoretical hypothesis to be just "a device for organizing the significant sentences of physical theory [the observation sentences]" and in doing so "to belittle molecules and their ilk, leaving common-sense bodies supreme." But, Quine continues,

> this invidious contrast is unwarranted. What are given in sensation are variformed and varicolored visual patches, varitextured and varitemperatured tactual feels, and an assortment of tones, tastes, smells, and other odds and ends; desks are no more found

[3] Quine 1951b, 134.

[4] Quine 1960a, reprinted in Quine 1966, 233-241. Quine tells us that this was originally written to be part of the introductory chapter of *Word and Object*, but "eventually superseded." But it was not superseded because of a change in view. Chapter 1 of *Word and Object* makes similar points.

> among these data than molecules . . . . In whatever sense the molecules in my desk are unreal and a figment of the imagination of the scientist, in that sense the desk itself is unreal and a figment of the imagination of the race.
>
> This double verdict of unreality leaves us nothing, evidently, but the raw sense data themselves.[5]

So far, this looks like a defense of phenomenalism, but Quine's twist was to take this line of argument as a reductio of the starting point. He continues: "Not only is the conclusion bizarre; it vitiates the very considerations that lead to it." The mistake is not the assumption that our sensory experience has "evidential priority," an assumption that Quine accepts, but rather the assumption that it has *ontological* priority. "Sense data are posits too. They are posits of psychological theory, but not on that account unreal."[6] The Quinean shift is to give a physicalist account of sense data (reflected in the rhetoric that he uses in *Word and Object* and other places to describe sense experience: "surface irritations," "impacts at our nerve endings," "ocular irradiation," etc.). Something like sense data are themselves posited as part of a theoretical account of the cognitive capacities of human beings, a theoretical account that is not a foundation for, but rather a derivative part of our overall theory of the physical world. So the agreement with Carnap's pragmatism is an agreement about epistemology: we accept an ontological framework, a physical theory stated in it, and the observation statements about ordinary things that it implies, because it provides an effective way of organizing and predicting our experience—experience that is described in terms of the resources that the theory gives us. But to do this is to accept that the theory is true ("subject to correction, but that goes without saying"[7]).

[5] Ibid., 237.
[6] Ibid., 238.
[7] Quine 1960b, 25.

While Quine blurs the line between pragmatic judgments about what it is fruitful to accept and theoretical judgments about what is true, there remains a sharp distinction that any realist must make between the *content* of a practical judgment that some statement should be accepted and the content of the statement itself. Any realist recognizes that the practical question, "should I believe that P?" cannot reasonably be answered in a way that conflicts with the answers one gives to the question, "is it true that P?" But he also recognizes that the *correct* answers to the two questions may diverge. (The recognition of this simple point is signaled in Quine's parenthetical remark, "subject to correction, but that goes without saying.") Carnap's position is anti-realist because he does not allow a distinction between the practical judgment that an ontological framework should be accepted and the theoretical judgment that it is a framework that is correct about what there is.

Quine distinguished his general approach to ontology from Carnap's in this way, but the method of doing philosophy that he promoted remains similar to Carnap's method of framework construction. Quine followed Carnap in proposing that the framework that is appropriate for expressing one's ontological and cognitive commitments is a constructed language, and he followed Carnap in proposing that the constructed languages should take the form of first-order quantification theory. They agree that the judgment about which frameworks of this kind to accept is a matter of decision, and that the decision should be based on the suitability of the framework for providing a fruitful way of systematizing our experience. Carnap observed that "the efficiency, fruitfulness, and simplicity of the use of the thing language may be among the decisive factors" motivating its acceptance, and he acknowledged that "the questions concerning these qualities are indeed of a theoretical nature."[8] These remarks show an implicit recognition of the

[8] Carnap 1950, 23–24.

undeniable fact that the process of framework construction and evaluation must itself be done in a context with its own language, and its own set of implicit commitments, but Carnap did not address the question, what framework are we working in when we assess the decisions about whether to accept a framework?

Quine does have an answer to this question, and it is one of the central features of his general philosophical approach: we start in the middle, with the language, theories, and beliefs that we find ourselves with at the point at which we begin the project of framework construction. "Our conceptual firsts are middle-sized middle-distanced objects, and our introduction to them and to everything comes midway in the cultural evolution of the race."[9] Quine's view was that the process of constructing and defending a formalized framework in this received informal framework needs to be seen as a bootstrap operation, a process, not of construction ex nihilo, but of revision and correction of the language and theory we begin with. This starting point includes both common-sense judgments about what is going on in our environment and received results of empirical science and mathematics. It also includes a recognition of the fact that we ourselves—the people who are theorizing and their practice of constructing languages and theories—are features of the reality that is the subject matter of our theorizing. As Quine put this point in the first chapter of *Word and Object*, "I propose to ponder . . . our scientific imaginings as activities within the world that we imagine."[10]

The framework we start with, Quine observed, does not come with a distinction between the part of what we accept that is based on invention or stipulation and the part that is discovery or hypothesis, and he argues that we don't need such a distinction for the process of revision and correction that he labels "regimentation," a process that involves both the construction of a language for saying

[9] Quine 1960b, 4–5.
[10] Ibid., 5.

in a perspicuous way what we believe and a statement in that language of those beliefs. The Quinean manifesto is that "the quest of a simplest, clearest overall pattern of canonical notation is not to be distinguished from a quest of ultimate categories, a limning of the most general traits of reality."[11] This manifesto is something of a rhetorical exaggeration, since Quine would acknowledge that some decisions about what canonical notion to adopt are more or less arbitrary and have little to do with ultimate categories, but he would say that the distinction is a matter of degree and must be made on a case-by-case basis.

There are at least two ways to think about Quine's project of regimentation, one more "metaphysical" than the other. On the one hand, one might think that the overall aim of the project is the construction of a comprehensive theory that is intended to represent the fundamental structure of reality. On this way of thinking, one's *ontology* is a domain of everything that exists, according to the theory one accepts, and the lexicon of the regimented language aims to provide all the resources one needs to describe that domain—to say everything there is to say at a fundamental level about what there is.[12] The supposition, from this perspective, is that the world has a metaphysical structure, and it is the task of regimentation to represent it. On the most uncompromising version of the fundamentalist view, any alternative to a correct theory of this kind that is not a notational variant of it, or of a part of it, is incorrect. A more liberal version of the fundamentalist view might grant that there could be alternative incommensurable fundamental theories that are equally comprehensive, but neither in conflict with each other nor inter-translatable.

A contrasting way of thinking about the project—a pluralist conception—is to think of regimentation as a process of clarifying various theories about particular subject matters, with no attempt

[11] Ibid., 161.
[12] See Sider 2011 for a development of this kind of metaphysical project.

to be comprehensive—no attempt to talk about everything at once. On this more modest view, there may be a general form common to the regimented languages that one constructs: a logical framework that is based on the idea that the way to represent a part of reality is by identifying an ontology and providing a lexicon for describing it, but this project need not purport to provide a single comprehensive ontology and lexicon. "On the whole the canonical systems of logical notation are best seen not as complete notations for discourse on special subjects, but as partial notation for discourse on all subjects."[13]

I think Quine is somewhat ambivalent about what kind of metaphysician he is. On the one hand, he says that he follows Carnap in being "no champion of traditional metaphysics," but he does acknowledge that the sense in which he uses the "crusty old word" *ontology* has its roots in traditional metaphysics,[14] and he famously remarked that one of the upshots of abandoning the two dogmas of empiricism is "a blurring of the supposed boundary between speculative metaphysics and natural science."[15] The florid rhetoric of the Quinean manifesto ("a quest for ultimate categories, a limning of the most general traits of reality") suggests the more ambitious interpretation, but Quine also defends what he calls a "*maxim of shallow analysis*" that fits with a more piecemeal project: "*expose no more logical structure than seems useful* for the deduction or other inquiry at hand."[16] When considering, in the last chapter of *Word and Object*, the task of "reconciling some limitedly useful lesser theory with a more sweeping theory," he advises that "we keep in mind also that knowledge normally develops in a multiplicity of theories, each with its limited utility and each, unless it harbors more danger than utility, with its internal consistency. These theories overlap very considerably, in their so-called logical

[13] Quine 1960b, 160.
[14] Quine 1951b, 127.
[15] Quine 1951a, 20.
[16] Quine 1960b, 160.

laws and in much else, but that they add up to an integrated and consistent whole is only a worthy ideal and happily not a prerequisite of scientific progress."[17]

Even if one's central concern is the "worthy ideal" of putting one's regimented theories together into "an integrated and consistent whole," doing this would be a further nontrivial task of regimentation in which questions about the relationship between the ontologies of the different partial theories need to be answered. The different theories, as Quine said, may "overlap considerably." In some cases, different theories may be talking in different ways about overlapping parts of reality, while in others they may concern separate and independent domains, and there will be hard cases where the relationship between the ontologies of different theories is more complicated. Agustin Rayo, in his defense of a deflationary and pluralistic view of ontology, uses what he calls "just is" statements to make a kind of claim about overlap of subject matter.[18] To Peter Van Inwagen's claim that there are no tables, even though there are things (atoms perhaps) arranged table-wise, Rayo counters that one might say "For there to be tables *just is* for there to be things arranged table-wise," meaning that a commitment to tables is not a further ontological commitment.[19] Of course, this is just a start of a story about the relationship between tables and the things that are arranged in certain ways; much more needs to be said about what it means to be arranged "table-wise" and about the identity of tables over time.

In easier cases, the ontology of one theory may be reduced to that of another, but even here there are complications. Consider Quine's discussion of "the ordered pair as a philosophical paradigm."[20]

[17] Ibid., 251.

[18] Rayo 2013.

[19] The claim is not that tables just are things arranged table-wise—that would be a straightforward identity claim. If there is an identity claim here, it is an identity of facts, and not of things.

[20] Quine 1960b, 257–262.

Are there ordered pairs, according to Quine? Yes and no. The theory of the ordered pair is in perfectly good order, but Quine said that we should "face the fact that 'ordered pair' is (pending added conventions) a defective noun, not at home in all the questions and answers in which we are accustomed to imbed terms at their full-fledged best."[21] He thought that the question, "Just what are ordered pairs?" should be dissolved rather than answered. We don't need ordered pairs in addition to sets (which we must acknowledge in any case) because we can use certain sets to play all the roles that ordered pairs play. If you like, you can say that ordered pairs are reducible to sets, so that it would be double counting to countenance them as further things. But (a familiar point) there are many incompatible but equally adequate ways of reducing ordered pairs to sets (of saying what ordered pairs *are*), so it would be arbitrary to endorse one over another. We might better say that "we can dispense with ordered pairs in any problematic sense in favor of certain clearer notions."[22]

Rayo might say that for there to be ordered pairs *just is* for there to be things arranged ordered-pair-wise, and here (in contrast with the case of things arranged table-wise) we can be precise about what it is for there to be things arranged ordered-pair-wise: It is for there to be an admissible three-place relational predicate, $R(x,y,z)$ meeting the following conditions: For any x and y, there is a unique z such that $R(x,y,z)$, and for any v and w, if $R(x,y,z)$ and $R(v,w,z)$, then $x = v$ and $y = w$. If our domain contains all sets, then there will be things arranged ordered-pair-wise, even though there is no one thing that can be said to be the ordered pair $\langle x,y \rangle$ for a given x and y.

It has been observed that there are use-mention issues with this way of putting the claim that there are things arranged ordered-pair-wise, and this is not just a pedantic point, but one

[21] Ibid., 258.
[22] Ibid., 260.

that helps to bring out a tension in Quine's general picture.[23] More carefully, what Quine might say is that to accept that there are things arranged ordered-pair-wise is to endorse a regimented language which admits a three-place relational predicate that meets the specified condition. If one accepts an ontology of sets, one thereby accepts the admissibility the predicates that are definable in set theory, and this implies that if set theory is true, then there are things arranged in this way. But since there are many different relational predicates meeting this condition that are definable in set theory, there is no answer to the question, "which of the things arranged ordered-pair-wise are the ones that are the ordered pairs?" without further stipulation that would be arbitrary.

The ordered pair, of course, is just one very simple example of a mathematical structure, and the same point could be made about any of them. Quine would also say, for the same reason, that "natural number" is a defective noun, but he would add that this is not a problem since there are things (sets) arranged natural-number-wise.

Whether or not one interprets Quine's project as aiming at a single most comprehensive metaphysical/scientific theory to which all else is ultimately reducible, the point I want to make is that the realism about ontology that distinguishes Quine's conception of a framework from Carnap's does not require that the ontology of a framework that one endorses make a claim to be comprehensive. It does not by itself require a commitment even to the intelligibility of a notion of absolute generality. What the realist is committed to a distinction between the questions, "do I, or should I, endorse a commitment to the ontology of this framework/theory?" and the question, "are the commitments correct?" To endorse a theory is to accept a positive answer to the second

[23] Thanks to Jeremy Goodman (p.c.) for raising this issue.

question, but not thereby to accept a claim about the comprehensiveness of the domain. The realist who endorses two different frameworks with different ontologies must acknowledge that we can always raise further questions about the relationship between the things in the different ontologies—about how the two frameworks fit together into a more inclusive framework. Such questions are asked in the pre-regimented framework in which we construct and assess our projects of regimentation—a framework that we perhaps never entirely leave behind as we refine our conceptions of what the world is like by constructing regimented theories. Even if each of two frameworks is in good order on its own, the project of merging them will be a further project of regimentation that will require answers to further questions, for example about the application of the predicates of one partial theory to the entities in the ontology of the other, and about the relations between the things in the two different domains.[24]

However one interprets Quine's actual metaphysical ambitions, my neo-Quinean picture will make only the more modest assumption that the endorsement of a regimented language/theory involves a commitment to the existence of things of the kind that the language purport to talk about, and it will remain noncommittal about the claim that there is or could be a comprehensive framework that describes the total universe at the most fundamental level.[25]

[24] The questions raised about the integration or calibration of different frameworks connect with questions that I discuss in the first and last chapters of Stalnaker 2008 about an absolute conception of reality. Bernard Williams had made the following observation: "Suppose A and B each claim to have some knowledge of the world. . . . A's and B's representations may well differ. If what they both have is knowledge, it seems to follow that there must be some coherent way of understanding why those representations differ, and how they are related to one another" (Williams 1978, 49). He goes on to say that "we must form a conception of the world that *contains* A and B and their representations." This suggests, to use a bit of Quinean jargon, a kind of semantic assent: we clarify a question by shifting from talk about substantive issues to talk of people who represent and their representations.

[25] The ideas expressed in the last few paragraphs were influenced by the arguments developed in Rayo 2013, and by discussion with him.

1.2. The character and scope of logic

The main point of the Quinean manifesto was to emphasize the continuity between the choice of a canonical notation and the decisions to accept substantive hypotheses, and it serves to highlight the contrast with Carnap's picture, but the fact that the process of regimentation does not presuppose an analytic/synthetic distinction does not mean that there are no distinctions to be drawn between different kinds of decisions that are involved in constructing a regimented language. Quine recognized that one role of a regimented language is to provide a common context for the perspicuous articulation of alternative hypotheses, and for rational discussion of and argument about the relative merits of those hypotheses. For these purposes we need a context that is neutral with respect to issues in contention, and this requires different levels of commitment in the acceptance of the regimented frameworks that the philosopher constructs. In particular, Quine argues for a distinction between the part of the project of regimentation that is the formulation of a *logic*, which can be shared by theorists with different substantive commitments, and the part that makes the substantive commitments. He acknowledged that his judgments about the shape that logic should take are ultimately justified, like the rest of his regimented languages/theories, by their fruitfulness in organizing and accounting for our overall experience, but he also took pains to separate the commitments of logic, as he understood it, from the more general commitments involved in accepting a particular ontology, and a lexicon for describing that ontology. He proposed to separate questions of meta-ontology (what is the appropriate framework for stating one's metaphysical commitments?) from questions of ontology (what commitments of this kind do I propose to make?) Questions of the first kind are answered by articulating a logical framework that requires, in application, the positing of a domain of things to be represented and a family of predicates with which to describe those things, but the logical framework is ontologically neutral in the sense that it is not committed to the existence of things of any particular kind. Claims of the second kind are answered by applying this logical framework by making posits of the kind required: a specification of a family of predicates for making claims about the kind of things that one proposes to talk about.

While these two kinds of questions are sharply distinguished, the neutrality claimed for logic is limited and specific; the claim is just that, while a regimented theory requires a commitment to a domain, which is the ontology of that theory, the logical framework itself makes no assumptions at all about what the things in the domain are like. It is decidedly not claimed by Quine that his theses about the specific shape that a logic should have, and about the nature of an ontological claim are theses that are known a priori, or beyond rational dispute. The epistemology remains thoroughly holistic. Quine recognizes that the commitments of logical theory that he endorses may be controversial, and his defense of those commitments is part of his defense of the whole package. For example, even though, as we will see, he argues that set theory is not a part of logic, he is happy to use set theory in his defense of his judgments about the form that logic should take. The soundness and completeness results for first-order predicate logic are (according to Quine) among the virtues of that regimentation of logic, and they play a role (acknowledged not to be decisive) in justifying the priority given to this logical theory. Even though those model-theoretic results are stated in a mathematical theory that is (according to his overall account of what there is) committed to sets, they still count as a point in favor of regimenting logic in that way. There is no vicious circularity here, since the strategy is to work from within. We start with the science and mathematics we find ourselves with, as well as with the apparent commitments of common sense. That is the pre-regimented theory that we are prepared to revise when we see problems, but we are also prepared to incorporate parts of it into our regimented theories when we think they will withstand scrutiny.

Just as the received context in which we begin does not come with a distinction between the commitments that are based on (implicit) stipulation and those that are based on discovery, so it does not necessarily come with a *theoretical* understanding of the semantics of the language we are using, or of the logic that is implicit in the reasoning we do in that language. What is given is a *practice*, which brings with it a competence in using the language we were brought up speaking to communicate and to reason. In formulating a regimented language, we rely on this competence, but not on a semantic theory of the language of the pre-regimented framework, and it was not part of Quine's project of regimentation to provide an empirical theory of the syntax or semantics of natural language.[26] Instead, the project was to construct a new language, with an explicitly articulated syntax and semantics, a syntax and semantics to be explained in the language of the pre-regimented framework with which we begin. This process of regimentation, Quine observed, is just a more general and systematic extension of the "opportunistic departure from ordinary language in a narrow sense [that is] part of ordinary linguistic behavior."[27] That is, the following is already true of the pre-regimented framework we find ourselves in: when things get complicated (e.g., in doing mathematics), we use variables and parentheses to facilitate cross-reference and to resolve potential scope ambiguities. Our competence with the natural language allows us to recognize scope distinctions even if we lack a general theoretical understanding of all the particular lexical and grammatical devices that are used to make those distinctions.

[26] This is not to deny that Quine might take the project of developing an empirical theory of the syntax and semantics of natural language to be a worthwhile scientific project. That is a separate issue. The point is just that the decisions about the structure of the regimented language are not to be confused with hypotheses about the workings of natural language. (I am not saying that Quine *did* take an interest in the project of developing an empirical theory of natural language. That is also a separate issue.)

[27] Quine 1960b, 157–158.

Instead of seeking such a theoretical understanding of the way the given natural language expresses logical relations and distinctions, the project was to construct a language with a simpler and more uniform system of devices for doing some of the things that we can do with that language. "Clearly it would be folly to burden a logical theory with quirks of usage that we can straighten. It is part of the strategy [of regimentation] to keep theory simple where we can, and then when we want to apply the theory to particular sentences of ordinary language, to transform those sentences into a 'canonical form' adapted to our theory."[28] In what sense are the new sentences *transformations* of the original sentences? There is, Quine says, no claim of synonymy. It is enough that we recognize that the new sentences in the canonical language are playing roughly the same role, in context, in representing how things are.

> By developing our logical theory strictly for sentences in a convenient canonical form we achieve the best division of labor: on the one hand there is theoretical deduction and on the other hand there is the work of paraphrasing ordinary language into the theory. The latter job is the less tidy of the two, but still it will usually present little difficulty to one familiar with the canonical notation.[29]

One might be skeptical that this kind of project could succeed since the new formal language is very different in structure from natural language. Can we really build a more precise language from within? I think Quine would say that the development of symbolic logic in the early twentieth century—itself already part of the received framework in which his philosophical project is to be assessed—demonstrated the fruitfulness of this kind of bootstrap operation. As he observed, "the artificial notation of logic is itself explained,

[28] Ibid., 158.
[29] Ibid., 159.

of course, in ordinary language,"[30] but it is clear that the linguistic and logical competence we started with was enough to give us the resources to understand the formal languages that logicians constructed, and it is also clear that the understanding of those formalized languages helps us to better understand and correct the reasoning that we do. (The familiar "bootstrap" terminology uses a manifestly impossible action *pulling yourself up by your own bootstraps* as a metaphor for a kind of thing—using given resources to improve those resources—that, somehow, can be done.)[31]

The division of labor that Quine proposed involves separating the project of articulating the abstract structure of reasoning (a structure that involves truth functions, predication and quantification) from the "less tidy" project of choosing and explaining a lexicon for the regimented language. The latter task provides resources for specifying particular subject matters that we reason about. The roles of the particular primitive predicates that we admit into the language are explained by paraphrasing them with the resources of the pre-regimented language in which we are working. The result of the first project sets up the *question* of ontological commitment; that is, it clarifies what form the answer to the question, "what is there?" should take. Interpreting the abstract formal language we are constructing requires the specification of a domain of entities that the quantifiers range over—the ontology of the language. These are the things to which (according to that theory/language) the predicates we admit are applied. We can say something substantive about what the commitments of the theory are only with the predicates that we introduce, so the act of making an ontological commitment cannot be separated from the decisions about what predicates we judge to be intelligible.

[30] Ibid., 159.

[31] See Verhaegh 2018, section 7.2 for a discussion of Quine's lectures for a course on the philosophy of language, given in 1953 as he was developing a draft of *Word and Object*. Quine's lecture notes elaborate on the role in the development of logical theory of paraphrase in natural language.

The upshot of this division of labor is that logic should be neutral about ontology, as well as about the lexicon that we admit to the language in which we describe the ontology. Quine had already taken pains, in his classic paper on ontology, published in 1948, to distinguish the question, (1) what are we doing when we make ontological claims? from the question, (2) what ontological claims are we in a position to make? He emphasized that his slogan, "to be is to be the value of a bound variable" is a "standard whereby to decide what the ontological commitments of a theory are. But the question what ontology actually to adopt still stands open."[32]

To help clarify Quine's views about the scope of logic—about the way he divides the labor—I will look at three questions he addresses: First, is identity theory a part of logic? Second, is set theory a part of logic? Third, how should logic treat proper names, or more generally, singular terms? A fourth question: is *modal* logic a part of logic? will come in for considerable discussion in later chapters. We postpone it here because it will be one of the central points at which our neo-Quinean theory diverges from Quine's.

Identity theory. The vocabulary of Quine's regimented language will contain predicates, variables, and what he calls "particles," which include operators, connectives, and quantifiers. The characterization of the logic specifies the semantic roles of the particles by giving compositional rules for each of them, but the interpretation of the predicates, on Quine's initial characterization, is not a part of the logic. The idea is that a logical truth will be defined as a sentence of the regimented language that will be true, however the predicates are interpreted, and whatever the ontology includes (holding just the interpretation of the particles fixed). The problem is that "this whole attitude toward logical truth is threatened . . . by the predicate '=' of identity," since this way of circumscribing logic implies that principles such as '$\forall x x=x$' are not logical truths, and this seems

[32] Quine 1948, 19.

wrong.[33] Quine argued that identity theory should be understood as a part of logic. I think this is the right call, and I agree with *some* of the reasons he gave for thinking that identity is a logical relation, but I don't think his way of reconciling his general characterization of the scope of logic with this judgment works. Quine argued that including identity theory in logic was compatible with his characterization of the scope of logic because the identity predicate (or at least "a serviceable facsimile" of it) is *definable* in logic as he had circumscribed it. He acknowledged that, strictly speaking, "we cannot define identity in terms of truth functions and quantification," but observed that we can define, in terms of a finite range of predicates, a notion of *indiscernibility* whose logic (in a language including just those predicates) was indistinguishable from the logic of identity. I think this is an unsatisfactory basis for including identity theory in logic for several reasons. First, Quine's "serviceable facsimile" of identity is itself not a logical notion, on his account of logic, since it depends, in application, on the particular range of predicates that are included in the overall regimented language. But second, it is in any case important to distinguish identity from any limited notion of indiscernibility. Only with a primitive identity relation can we say that there is an x and a y such that $x \neq y$, even though x is indiscernible from y (with respect to all the predicates other than identity in a given regimented language).

The real basis for distinguishing logical from non-logical notions is not the grammatical distinction between predicates and particles, but the distinction between notions that are topic-neutral and those that depend, for their interpretation, on a particular ontology, and on their place in a particular range of predicates (those that apply to things in that particular ontology). Whatever one is talking about, or purports to be talking about—numbers, persons, quarks symphonies, sets, fictional characters, ghosts, propositions, functions, attributes, or whatever—the relation of identity will be

[33] Quine 1986, 61.

applicable to that ontology, and in fact identity is a particularly important relation (for Quine) for the critical assessment of any ontological claims. As Quine put the point, "Another respect in which identity theory seems more like logic is its universality: it treats all objects impartially."[34]

The labels "topic neutral" and "universal" are not quite right for saying clearly what is distinctive about identity. As Jeremy Goodman pointed out to me, a predicate like "interesting" is topic neutral in the sense that any putative thing (quarks, symphonies, propositions, ghosts, etc.) might be described as interesting without making some kind of category mistake, but no one would be tempted to think of "interesting" as a logical predicate. And identity is of course not universal in the usual sense that it is true of every pair of things in the domain. More carefully stated, the point is that for any putative domain, the identity predicate must be among the predicates that apply to the things in that domain. Even if some quarks or natural numbers might be judged by some people to be more interesting than others, the physicists or number theorists might reasonably say that their theory does not admit this predicate and does admit the question whether it applies to the things they are talking about. (The endorsement of an ontology and the family of predicates that comes with it includes the endorsement of the questions that it raises). Whatever the ontology you are committed to, you can't get away with saying that identity is not in the range of predicates that are admissible for describing the things in it.

Set theory. The contrast between the question whether identity theory is a part of logic and the question whether set theory is a part of logic helps to reinforce the point about the way in which logic (according to Quine) is neutral about ontology. In the case of set theory, Quine reached the opposite conclusion from the conclusion about the status of identity theory, and his reason was that set theory is a theory about a particular kind of object, characterized

[34] Ibid., 62.

with distinctive predicates, such as membership and inclusion, that apply distinctively to sets.[35] Pure set theory can be regimented in a first-order language that has just one primitive predicate—the membership relation—but one cannot understand this relation independently of the acceptance of a theory that is committed to a kind of entity that can stand in this relation. (The general theory puts no restrictions on the kind of thing that might be a *member* of a set, but it does presuppose that only a particular kind of thing can *have* members, and it would be consistent with logic to deny the existence of things of that kind.) Quine himself accepted (reluctantly) the existence of sets, but he thought that the tendency to think of set theory as a part of logic distorted the character of the question whether set theory should be accepted, as well as the question of what commitments should be involved in the theoretical regimentation of logic. The notion of predication is central to logic, but it is a mistake to try to explain monadic predication as a statement of a relation between two things. "The tendency to see set theory as logic has depended early and late on overestimating the kinship between membership and predication." We can agree that a monadic predicate has an extension, which can be represented (in a set-theoretic model of the regimented language), by a set (assuming there are sets). The predicate will be true of an object in the model if and only if the object is a member of the set that represents the extension of the predicate. But this cannot be a general explanation of predication, since the membership claim is itself a statement that applies a binary predicate to the object and the set that represents the extension of the monadic predicate. We are not entitled to assume that the intelligibility of predication requires that we think of the predicate as something that denotes, or is associated with, an entity of some kind (either an attribute or a set that is its extension).

The conclusion that set theory is not a part of logic is a rejection of logicism—the thesis that all of mathematics can be reduced to

[35] Ibid., 64–68.

logic. And it shows that Quine's attempt to distinguish the truths of logic from the more substantive truths stated in a regimented language does not either presuppose or yield anything like an analytic/synthetic distinction. Most mathematical truths (as well as all the theses involving empirical predicates that Carnap would have called "semantic rules") count as substantive, nonlogical truths, since they involve or presuppose substantive ontological commitments.

But as noted above, the question of the logical status of set theory is complicated, and one might hold (and Quine might agree) that while set theory is not a part of logic itself, it is an essential part of the theory of logic, which includes a semantics, or model theory for the language. Logical truths are sentences that are true in all models, where models are set theoretic structures. But even if a commitment to sets is essential to a theoretical understanding of logic, it won't follow that sentences affirming the existence of sets are thereby logical truths.

Singular terms. The vocabulary of Quine's regimented languages includes logical particles, variables, and predicates. What about names, or more generally, singular terms that denote individuals, and that combine with predicates to determine sentences that say something about the individual denoted? Quine pointedly excludes singular terms from his regimented languages, even though they are a part of the unregimented language with which we begin. The only expressions that represent the things that predicates are predicated *of* are variables, and these don't refer to particular individuals, but are placeholders for the determination of complex predicates, and for the representation of quantified claims. That is, Quine's proposed regimented language takes the form of a *pure* predicate logic: the language is one in which all closed sentences are purely general claims saying that some things (or all things) meet a certain general condition, or are truth-functional combinations of claims of this kind.[36] This raises two questions, both of which

[36] Quine's strategy for eliminating singular terms is spelled out in section 37 of Quine 1960b, 170–180.

Quine addresses: First, what is the motive for excluding names from the lexicon of a regimented language? Second, how can this be done without limiting the expressive power of the regimented language in an unreasonable way?

First, why does Quine want to ban names? The problem that names raise for the project of circumscribing the scope of logic might be put this way: In a standard first-order language with names such as '*s*' for Socrates, one can infer 'F*s*' from '$\forall x Fx$', for any predicate 'F', or more generally, one can infer $\phi^{s}/_{x}$ from $\forall x\phi$, by universal instantiation. This implies that since '$\forall x \exists y x=y$' is a logical truth of the theory of identity, we can apply this universal instantiation rule to get the conclusion, '$\exists y s=y$', which says that Socrates exists. Every logical consequence of a logical truth is itself a logical truth, but while the conclusion that Socrates exists is true, given the tenseless notion of existence, it should not be a *logical* truth, given Quine's conception of logic as ontologically neutral. As the Quinean manifesto says, one cannot separate the task of constructing a language from the decision to accept certain empirical commitments, but one should, Quine thought, be able to separate the task of articulating a *logic* from the tasks of specifying a lexicon of the language in which particular ontological commitments are expressed. Both names and predicates have substantive presuppositions, but Quine's project assumes that we can quarantine the commitments of logic from the more general commitments involved in the choice of a specific lexicon. This does seem to be possible for *predicates*, since the admissibility of a predicate to the lexicon does not say anything about its extension. (In particular, it does not imply that it has a nonempty extension.) But a name wears its factual presupposition on its face: if we allow names into the lexicon at all, then a name should be admissible if and only if it has a referent; that seems to be an unavoidable requirement, at least if our logic validates a rule of universal instantiation, and that rule seems to be essential to the logic of quantification. One might try to keep the ontological neutrality of logic by qualifying the rule of universal instantiation, but that would raise a number of problems. I will consider that alternative response after looking at Quine's answer to the second question—at his reasons for thinking that names can be eliminated without any real loss of expressive power.

We can do without names, Quine argued, by simulating them with predicates that are stipulated to apply to just one specified individual. So for example, the name 'Socrates' in the unregimented language with which we begin allows us to pick out a particular individual. Instead of introducing into the regimented language a name 's' for Socrates, we could introduce a monadic predicate that is stipulated to apply to Socrates, and to no one else. For anything that can be said in the pre-regimented language with the name 'Socrates', we can say a corresponding thing in the regimented language with the new predicate. The pre-regimented statement 'Socrates is wise', for example, can be replaced in the regimented language by '$\exists x(Sx \wedge Wx)$', where 'S' is a primitive predicate 'is identical to Socrates' and 'W' is the predicate 'is wise'. The statement that Socrates exists, which with the name would have been formalized as $\exists xs = x$, is instead formalized as $\exists xSx$, which is not a logical truth, on Quine's way of delimiting the truths of logic.

The point of this maneuver was not to reduce a predicate that conceptually involves a particular individual to something purely qualitative, since the individual himself is essential to the intuitive explanation of the predicate, and so to the judgment that the predicate being introduced to the regimented language is admissible. One might say that the admissibility of a predicate whose role is explained in this way still presupposes the existence of the referent, but what is accomplished by this maneuver is to separate this commitment from the commitments of the logic. All the logic tells us about the predicate 'S' is that it is a monadic predicate, and a statement involving that predicate will be a logical truth only if every statement with another predicate uniformly substituted for 'S' is true. Since it is not a presupposition of the admissibility of a predicate that it has a nonempty extension, '$\exists xSx$' will not be a truth of logic.

This maneuver works, and it reinforces the picture of the process of regimentation as a process of revision and correction of the language we begin with. We have the capacity, in that pre-regimented language to pick out particular individuals. In cases where we have succeeded in identifying an individual, we can stipulate that a predicate of the regimented language shall have just that individual as its extension. Of course the pre-regimented language may contain names that misfire, as did the introduction of the planet name 'Vulcan', stipulated to apply to a planet that was hypothesized to explain some of the features of the orbit of the planet Mercury. It turned out that there was no such planet, so the empirical presuppositions that were made in the attempt to specify the extension of the name were not met. Therefore, if one attempted to introduce a predicate 'V' into the regimented language in this way, it would misfire in the same way that the name did in the pre-regimented language. One could still use a description, "is a planet that plays such and such a role in the explanation of the orbit of Mercury" to specify a monadic predicate, and this predicate might be admissible to the language for an astronomical theory, even though the predicate in fact had an empty extension.

Even if Quine's choice to eliminate names succeeds in allowing the regimented language to talk about particular things in the domain, as I think it does, one might ask whether the maneuver is necessary. It is worth considering an alternative choice about how to construct a regimented language. The alternative allows names, and involves a small change in the classical logic of quantification, but the resulting language has essentially the same expressive power as the language that results from Quine's proposal.

The problem about names, for Quine, was that the ontological commitment implicit in a name was built into the logic, and this is not compatible with the assumption that logic should be ontologically neutral. Avoiding the problem by banning names masks the fact that the standard first-order logic with names *does* build the presupposition that names have reference into the logic. One might think that the right solution, or at least an alternative solution, to Quine's problem is to recognize that the standard first-order logic for a language with names is too strong. We can agree that it is a mistake to build the ontological commitment presupposed by the introduction of a name into the logic, but one might think that the mistake should be corrected, not by eliminating names, but by moving to a free logic in which the inference from a universally quantified statement to an instantiation containing a name, to be valid, will require that the existential presupposition be made explicit as a premise.

Quine and others resisted the move to a free logic since it seems to require distracting complications, including a worry about bivalence, or a problematic distinction between positive and negative predicates. But properly done, free logic is straightforward. It is compatible with bivalence, and it requires no commitments that are not implicit in the standard first-order logic, and only minimal changes to logical principles.

The apparent problem with allowing empty names is this: Take our example of 'Vulcan' (a name for which the empirical presupposition made in attempting to fix its reference was false). What do we say about a statement such as 'Vulcan is a planet', or 'Vulcan is closer to the sun than Mercury'? They aren't true, so to preserve bivalence, we must say they are false. But then 'Vulcan is not a planet' will be true. Do we then need to distinguish between positive predicates, such as 'being a planet' and negative predicates such as 'not being a planet'? The answer is no, we don't need any such distinction. What we need is just a scope distinction between negation as a predicate operator and negation as a sentence operator, and variable binding gives us that distinction for free. Since it is false that Vulcan is a planet, it is true that it is not the case that Vulcan is a planet. But it is also false the Vulcan is a non-planet, where the predicate 'being a non-planet' is definable in terms of the predicate 'being a planet'.

In the traditional formulation of quantification theory, the basic variable-binding operators are quantifiers, which combine in one operator two distinguishable functions: first, the implicit formation of a complex predicate and, second, an expression of generality. As is well known and not controversial, one could make the formation of a complex predicate explicit, using exactly the same Tarskian variable-binding machinery to specify the semantics for an operator that takes open sentences to monadic predicates. Instead of this semantic rule

$$[\![\forall x\phi]\!]^{s} = 1 \text{ iff } [\![\phi]\!]^{s[d/x]} = 1 \text{ for all } d \in \mathrm{D},$$

we have the rule

$$[\![\lambda x\phi]\!]^{s} = \{d \in \mathrm{D}: [\![\phi]\!]^{s[d/x]} = 1\}$$[37]

With this notation, we can distinguish '$\lambda x{\sim}\mathrm{F}x(t)$' from '${\sim}\lambda x\mathrm{F}x(t)$'. If '$t$' is an empty name, then for any interpretation of 'F', the former will be false, and the latter true. So the distinction between the false statement that Vulcan is a non-planet and the true statement that it is not the case that Vulcan is a planet is a simple scope distinction.

This alternative way of regimenting predicate logic, with complex predicates represented explicitly, should be unproblematic from a Quinean point of view, and it is also conceptually clearer than the traditional formulation. In the traditional formulation, complex

[37] Given Quine's concerns about confusing referring expressions with predicates, it is important that I be clear about the notation I am using. The lambda notation was originally introduced to distinguish names for functions from functional expressions whose denotation (for a given argument) is the value of the function. So, for example, 'x+3' denotes 7 when $x = 4$, while '$\lambda x(x+3)$' denotes the plus 3 function. But the lambda notion has come to be used for the formation of complex monadic predicates, and that is how I am using it. An expression of this form is a complex monadic predicate with exactly the same semantic role as a primitive monadic predicate. It is not a name of a function, or of anything else. No ontological commitments beyond those Quine would be happy with are smuggled in with this notational variation of the traditional formulation of predicate logic.

predicates are essential to understanding quantified sentences, but they are only implicitly represented in them. To say that all penguins are flightless, ('$\forall x(Px \rightarrow {\sim}Fx)$'), is to say that everything satisfies the complex predicate 'being flightless if a penguin' ('$\lambda x(Px \rightarrow {\sim}Fx)$'), but no predicate of this form occurs in the regimented sentence. The fact that complex predicates are represented only implicitly forces one to treat some sentences as substitution instances of others (with implicit complex predicates substituted for simple predicates), even though they are not substitution instances in a strict sense. Strictly, a substitution instance is the result of substituting a complex expression of a certain type uniformly for a simple expression of that type. It is reasonable to assume that any substitution instance of a logically valid sentence should itself be valid, since the simple predicate letters in the language represent arbitrary expressions of the same type. But it is not obvious that the looser substitution rule should preserve validity. For example, consider the sentence '$(Ft \rightarrow \exists xFx)$', which has some claim to be a logical truth (even in a language that allows empty names) since only existing things are candidates to satisfy a predicate. '$({\sim}Gt \rightarrow \exists x{\sim}Gx)$' is treated as a substitution instance of this sentence, implicitly taking '${\sim}G\_$' to be the result of substituting a complex predicate for the simple predicate F. The substitution presupposes that '${\sim}Gt$' will be equivalent to '$\lambda x{\sim}Gx(t)$', which, as we have seen, a free logic will deny. A scope equivocation is built into the traditional notation. Given the usual assumptions of predicate logic, where it is presupposed that all names refer, the expressions with different scopes are logically equivalent, but even if one stays with the classical assumptions, the equivalence should not be built into the notation. So I think the formulation with explicit complex predicates is independently motivated.

While this notational variant of first-order extensional predicate logic should be unproblematic from a Quinean point of view, it does suggest a slightly different way of thinking about Quine's meta-ontology. If variable binding is essentially the formation of complex predicates, then the slogan "to be is to be a value of a bound variable" might be paraphrased as "to be is to be a subject of predication." The commitment of a Quinean regimented theory has two interconnected components: the positing of an ontology, and the decision to admit a certain range of predicates. They are interconnected since there is no way to say what domain one is ontological commitment to except by using the predicates that may be used to distinguish the things in it from each other. Since the process of constructing and assessing a regimented language takes place in the context we find ourselves in, we begin with the predicates of ordinary language and the language of received science that we take ourselves to have some understanding of, sharpening them by looking critically at the relationships that seem to hold between them. In doing this, we are also clarifying the nature of the ontological commitments we are endorsing.

The addition of complex predicates does not increase the expressive power of the first-order language with identity, either with a free logic or with a logic with the usual existential presuppositions built in. This is because one can get the effect of a complex predicate by using a quantifier and identity. In the traditional formulation, one could take $\lambda x{\sim}\mathrm{F}x(t)$ to be an abbreviation for $\exists x({\sim}\mathrm{F}x \wedge x{=}t)$. A sentence of this form makes a general statement, and then subtracts out the generality with an identity clause. The result is logically equivalent, but distorts the simple structure of the complex predicate.

A predicate-forming operator is the only variable-binding operator we need. With it, we can, without loss of expressive power, treat quantifiers as simple, non-variable-binding operators on monadic predicates: Our regimented language could have a formation rule like this:

If 'F' is a monadic predicate (simple or complex), then '∀F' is a sentence.

And a semantic rule like this:

$$[\![\forall \mathrm{F}]\!] \text{ is true iff } [\![\mathrm{F}]\!] = \mathrm{D}.$$

A quantified sentence such as $\forall\lambda x(Fx \rightarrow Gx)$ will then look familiar, except for the 'λ' and will be logically equivalent to the traditional formulation, without the 'λ'. It will, however, have a different grammatical structure, separating the two different jobs that variable-binding quantifiers are doing in the traditional formulation.

The fact that the language of predicate logic was originally formulated (in Russell and Whitehead's *Principia Mathematica*), with quantifiers as the variable-binding operators was a quirk of its history, facilitated by the fact that the scope distinctions it ignored were, given its assumptions and limited expressive power, not semantically significant. As we have observed, Quine remarked in setting out his program of regimentation that there was no point in building the quirks of natural language into our regimented language. The same can be said for the quirks of the traditional formulation of the language of first-order predicate logic.[38]

I have presented the move to a free logic as an alternative to Quine's maneuver of eliminating names and replacing them with predicates, but these two ways of regimenting the use of names and other singular terms are, in a sense, equivalent. The monadic predicates with which Quine replaces names will have an extension that is either unique or empty. The terms with which the free logician regiments singular terms will also have either unique or empty extension. In both cases the regimented replacement of a simple predication ('Socrates is wise' or 'Vulcan is a planet') will be a sentence that is true if and only if the term that replaces the singular term of the pre-regimented language has a nonempty extension that is a part of the extension of the predicate (is wise, or is

[38] The proposal to formulate the logic with variable binding used to form complex predicates and to treat quantifiers as non-variable-binding operators on predicates is independent of the issues about names. However one treats names, I think this proposal is well motivated. This proposal is also independent of the proposal made in free logic that we should allow for the empty domain.

a planet). On neither proposal is there a presupposition built into the logic that the term that replaces the ordinary name has a nonempty extension. On both proposals, *one* way to introduce an expression of the kind that replaces names is this: one refers in the pre-regimented language to an individual and stipulates that the expression introduced into the regimented language shall have that individual as its extension. But on both accounts one also *might* allow the regimented term to be stipulated to have as its extension whatever it is, if anything, that satisfies a certain condition (presupposing that at most one thing satisfies the condition). There are thus two ways to think about the regimentation of a name like 'Vulcan': (1) it is a failed attempt to use a particular individual, hypothesized to exist, to fix the reference of a term, or (2) it is a term whose extension is explained in the second way, as whatever fits the description (if anything) of being the planet that accounts in such and such a way for the perturbations of the planet Mercury. In the former case, it is presupposed that the introduced term has a nonempty extension, but this presupposition is imposed, not by the logic, but by the constraints on admissibility for nonlogical terms. The term is inadmissible to the lexicon of the regimented language for factual reasons. In the latter case, the introduced term would be admissible, but existence will not be presupposed by its admission to the regimented language.

The change to the logic that is proposed in this alternative to Quine's decision to eliminate names is an isolated change: it concerns the role that singular terms play, but it has no effect on the pure logic of quantification, which remains exactly as it would be if we followed Quine's way of using predicates to permit one to talk, in the regimented language, about particular individuals.

In the traditional formulation of predicate logic, an ingenious and elaborate paraphrase (the Russellian analysis) is necessary to allow for definite descriptions in the regimented language, but with the complex-predicate formulation, one can treat definite descriptions as ordinary singular terms with a semantics that makes sentences containing definite descriptions logically equivalent to those that result from the Russellian analysis. That is, one can introduce a definite description operator, 'I' on predicates,[39] so that if F is a monadic predicate (simple or complex), then 'IF' is a singular term denoting the unique thing that satisfies F, if there is one, and nothing otherwise.

The general point of Quine's attempt to circumscribe logic is something like this: The overall project of regimentation involves the formulation of substantive theory, including commitment to particular ontologies, and to the intelligibility of the lexicons that are used to characterize those ontologies. But a part of this project of regimentation is the choice of the general shape that a regimented language/theory should take, and it is part of the strategy to try to characterize in a precise way the questions to be asked and answered in the construction of a regimented language, and to do so, as far as this is possible, in a way that is independent of the particular answers to the questions that one gives and defends. It is not a foregone conclusion that this division of labor can be accomplished, and it is acknowledged that the articulation of the general shape of the framework for stating our theories about what the world is like is itself a part of the overall project, done from within, and starting with the language and theory we find ourselves with. But since we want a framework not only for stating clearly our present settled views but also for engaging in rational debate with others who do not share those views, and for formulating unanswered further questions about what our views should be, we have reason to try to distinguish different levels of commitment, and the circumscription of the commitments of logic is a part of that aim.

[39] Russell used an upside-down lowercase iota for this symbol. For typographical convenience, I follow David Kaplan in using an upside-down uppercase "I."

1.3. Summary of the Quinean picture

I will conclude the chapter by highlighting those features of the Quinean philosophical project that I will draw on in carrying out my own neo-Quinean project, which will depart from Quine's project in significant ways.

No transcendent perspective. The central idea that guides the development of Quine's philosophical projects is that there is no "foundation for scientific certainty firmer than the scientific method itself," and the scientific method itself is something to be described and justified from within, as a part of the substantive philosophical/scientific picture of the world that we are prepared to endorse. (How many times did Quine invoke Otto Neurath's metaphor of rebuilding a ship at sea?) But the rejection of an a priori foundation for philosophy raises a general problem: how can we, working with the unsystematic, ill-understood, disunified cluster of languages and theories that are available to us, construct and justify clearer, more precise, and more accurate pictures of the world we find ourselves in? Most of the features of Quine's philosophical methodology are parts of a strategy for answering this question.

Regimentation. The form that Quine's philosophical projects were to take follow closely the method that Carnap and others in the logical empiricist movement called "explication" and which Quine relabeled "regimentation": the construction of a formal language for the representation of our mathematical and empirical scientific theories. But the rejection of a transcendent perspective focused attention on the pre-regimented framework in which we develop and assess our constructed languages. As the new label suggests, the process is a matter of assessing and revising the received languages and theories, explaining both the semantic structure of the constructed languages and their vocabulary in terms of our pre-theoretical understanding of the natural language we begin with. But we should not think of the process as one of simply *explicating* the meanings of the terms of the pre-regimented language that are used to explain them. Quine agreed with Carnap and the logical empiricists that this process involves decision about how to talk, and not just reports on how we do talk. But the decisions have presuppositions that make substantive commitments.

Pragmatism and realism. Quine, like Carnap, was a pragmatist, which meant that the critical assessment of the frameworks we construct should be based on their fruitfulness for explaining our experience and for finding our way around in the world. But unlike Carnap, Quine was a realist about the kind of commitments that we make. "Physical objects are real, right down to the most hypothetical of particles, though this recognition of them is subject, like all of science, to correction. I can hold this ontological line of naïve and unregenerate realism, and at the same time I can hail man as largely the author rather than discoverer of truth." We invent and endorse theories, but truth is a matter of whether the world answers to the theories we invent and endorse. It was not that Carnap regarded himself as an anti-realist; he held that realism is metaphysics, and so meaningless rather than false. His point was that judgments about what is real or true can be made only within a framework. But Quine's point was that all our judgments, including choices about what frameworks to endorse, are made within a framework—beginning with the framework we are in when we start theorizing. "We are always talking within our going system when we attribute truth; we cannot talk otherwise."[40]

Meta-ontology. Among the judgments that we make in the pre-analytic framework we begin with are judgments about what form a regimented theories should take, and it is an important feature of Quine's methodology that he separates questions about this from questions about which specific regimented theories to endorse; that is, he distinguishes meta-ontology from ontology. The point of this separation was not that questions of meta-ontology are asked from

[40] Quine 1975, 33–34.

an a priori transcendent perspective, or that their answers should be regarded as beyond dispute, or as not subject to revision. His meta-ontological presuppositions, including presuppositions about what part of a regimented theory should count as logic, are part of the overall set of commitments about the kind of world we live in, and the languages that are apt for describing it, an account that is to be judged by its fruitfulness in explaining our experience and finding our way around. The rationale for the separation is just that reasonable methodology should be expected to involve some division of labor, and not ask us to swallow a package of commitments whole, or not at all.

Ontology and predication. The specific form that a regimented theory should take, according to Quine, is a first-order extensional predicate logic with identity. To endorse a theory of this form is to undertake a commitment to the existence of a domain of entities and to the admissibility of a family of predicates for characterizing the things in the domain. The two commitments (to an ontology and to a family of admissible predicates) are interdependent, since there is no way to understand the ontology except as things that are described in certain terms. So the primitive notions of any theory of this form are the notions of existence and predication, notions that are explained informally in the natural language that we begin with. We use our competence with this language to identify particular domains of things we think are to be found in our world, and the predicates for talking about them that we think we understand, at least after some critical examination and clarification. Then in that pre-regimented language, we define a constructed language/theory of the given form. The meta-ontology aims to be ontologically neutral in the sense that it makes no assumptions about the kind of thing that might be in a domain of a theory we might be prepared to endorse. To be is to be a subject of predication. The assumption of the meta-ontology is that for anything at all that we are prepared to believe exists, we can have a first-order theory of things of that kind. One point of this way of dividing the labor of theory construction is that it helps us separate commitments about what there is in the world to talk about from questions about the language we use to talk about those things. The rejection of the analytic/synthetic distinction is not a rejection of any separation of metaphysical and semantic questions; it is just a recognition of the fact that the task of separating questions about the world from questions about how to talk about the world is a substantive theoretical task, and not one to be settled prior to developing substantive theory. Quine's notorious obsession with use/mention conflations is motivated by a concern with the important of this difference.

Pluralism, in practice. Even the general presuppositions about the shape of a regimented language/theory are subject to correction and revision. Regimentation is a dynamic process, taking place piecemeal in an evolving background language that is being enriched by the addition of parts that are regimented in various ways. Some Quineans may have a utopian vision of a single regimented fundamental theory at the end of inquiry, one that supersedes the evolving language in which we work, but that is not an essential part of the story (and in my view, not a plausible part of the story).

My own project will start with these methodological assumptions, though some of them will change as we proceed. In the next chapter, I will begin with an outline of how my general argument will go.

2
Propositions

2.1. Overview

I start this chapter with a sketch of the dialectic of the main overall argument that I will be developing in this book, emphasizing the role that the methodology discussed in Chapter 1 will play in it. As noted in the introduction, Quine considered and rejected the hypothesis that there are such things as propositions. The first move of my argument will be to contest this conclusion, at least for one specific way of regimenting the concept of a proposition. After spelling out a specific theory of propositions in a first-order extensional language (section 2.2), I will argue in section 2.3 that Quine's specific reasons for rejecting propositions do not apply to the notion of proposition that I am defending, that this notion meets the conditions that Quine requires for the clarity of an ontological commitment, and furthermore, that it is in fact a notion with which he expressed sympathy.

The next move of my argument, taken in section 2.4, is the simple observation that if there are propositions of the kind that my theory endorses, then it is plausible to assume that the sentences of the kind of regimented theory that Quine aims to construct, whatever its specific ontology, should *express* propositions. That is, we should require that the form of a regimented theory should determine not just a truth-value but a proposition for each of the sentences of the language, and this means that we need a logic with a more demanding compositional semantics—specifically, a modal logic. Quine famously rejected modal logic, arguing that its foundation rested on a use-mention confusion, and I agree that Quine's

Propositions. Robert Stalnaker, Oxford University Press. © Oxford University Press 2023.
DOI: 10.1093/oso/9780197647035.003.0003

objections were well taken, as applied to the way modal operators were explained by C. I. Lewis in his development of modal logic in the first part of the twentieth century. But I argue (in section 2.5) that our theory of propositions can provide a foundation for modal logic that is not subject to Quine's objections.

At this point, we are not only making a specific ontological commitment that Quine was not prepared to make but also modifying his general methodological assumption about the character of logic, and about the form that a regimented theory should take. We retain, however, the kind of ontological neutrality that motivated Quine's characterization of logic. A first-order modal logic, like first-order extensional logic, makes no presuppositions about the nature of the entities that are in its domains. The hypothesis that there are propositions plays a central role in the *motivation* for the acceptance of modal logic, and in the model theory for that logic, just as the acceptance of the hypothesis that there are sets played a role in the model theory that was part of Quine's reason for giving a central place to first-order logic, but that does not imply that propositions (or sets) will be among the things in the domain of each regimented theory that we are prepared to endorse.

The theory of propositions provides the resources for a defense of modal logic—for a compositional semantics for the modal operators as well as the truth-functional operators—but it also raises new questions about the interpretations of the *predicates* of a regimented language—both general questions about criteria for the admissibility of predicates to a regimented language, and specific questions about the particular predicates of our theory of propositions. In Chapter 3, I will argue that the general questions raised by the enrichment of the logic are questions the Quinean project faces in any case, and that the theory of propositions provides some resources that help to sharpen and answer them.

While our account argues for an enrichment of logic, the logic that we use at this point in the development of the argument remains a first-order logic, and the only ontological commitment beyond those made by Quine that we have defended is to propositions. Chapter 4 pauses to spell out in detail a model theory for first-order modal logic, with an axiomatization that is sound and complete with respect to that model theory. Then in section 4.3 of that more technical chapter, I spell out the application of the model theory to a particular regimented language where the domain of the theory is a set of propositions. While the model theory involves (in the meta-language) set theoretic entities that are the semantic values of predicates as well as of sentences, the regimented languages we consider at this point are all first-order, and none of them have entities in their domains that correspond to predicates or to sentences. The clarification, in Chapter 3, of admissibility conditions for predicates provide some resources that will help to defend an extension of the ontological commitment to include properties and relations—entities that are expressed by the predicates—in addition to propositions, but at this stage of the argument we have refrained from making this further commitment. Chapter 5 will begin by entertaining the proposal that we should extend our ontology in this way. It is important for the dialectic of the overall argument that at this point we are making no proposal about the form of a language that is appropriate for talking about what there is; we are just proposing an ontological hypothesis—that there exist properties and relations. But unlike the hypothesis that there exist propositions, the hypothesis that there are properties and relations corresponding to the predicates of a regimented language ramifies. To be, on the Quinean meta-ontology, is to be a subject of predication, so if there are properties and relations corresponding to the predicates of a first-order language, there must be a family of predicates applying to those entities—ways of describing them—and these, too, on our hypothesis, will correspond to properties and relations, and that start us off up a hierarchy of properties and relations. The admissibility of predicates, on the Quinean picture, is conceptually tied to the commitment to an ontology: predicates are admissible if and only if they are apt for describing things in the given ontology. So we should expect that the ontological hierarchy of properties and relations will be typed in a way that corresponds to the particular domains of entities that are candidates to be the things that are exemplified by a given property or relation. The type theory that is appropriate for characterizing the proposed hierarchy of properties and relations looks a lot like the type theory of higher-order modal logic, where the things that are typed are expressions, typed by the entities that are their semantic values, and quantifiers and variables, typed by the domains over which the quantifiers range. This kind of higher-order modal logic seems to be an appropriate framework to develop a general theory for the ontological commitment we are considering, and a framework of this kind, developed by Richard Montague and modified by Daniel Gallin, has been used in recent years to develop a metaphysical theory of attributes and propositions.[1] Quine rejected higher-order logic, even at the extensional level. He acknowledged the coherence of such logics, but he warned that they posed a danger of equivocation, and of the blurring of the line between talk of the things in an ontology and talk of the language we use to theorize about those things. I think Quine's qualms about higher-order logic are well motivated, and I will take his warnings seriously; however, I will argue that with care we can avoid the equivocations, and that the higher-order theory, properly understood, provides a useful framework for formulating a theory about the structural relations between the entities of all the different types. But my Quinean interpretation of higher-order modal logic will differ in significant ways from the interpretation presupposed by others who are using this framework to develop a metaphysical theory in which modality plays a central role. I will consider, briefly, some aspects of alternative, anti-Quinean interpretations, but I will leave any detailed discussion of them for another occasion.

[1] See Williamson 2013.

In section 5.3 of Chapter 5, I will consider whether one might formulate a theory of the rich typed ontology of propositions, properties, and relations in a single first-order regimented theory. Such a formulation would raise questions that are not settled by the higher-order theory, and it faces a threat of paradox not faced by the higher-order formulation.

Chapter 6 reflects on the relation between the model theory for a modal language and the reality that we are aiming to model.

2.2. A regimented theory of propositions

The theory of propositions that I will defend will be formulated (at least initially) in first-order extensional logic with identity. It is a simple and familiar theory—the structure is an atomic and complete Boolean algebra, though I will axiomatize the theory in a nonstandard way. The ontology (the domain of the first-order language in which we formulate the theory) will include propositions and sets of them. While we have some sets in our domain, we will need only the most elementary parts of set theory, since the only sets in the domain of our theory will be sets of propositions.[2] Our theory is sensitive to Quine's concern with identity conditions. For a kind of thing to be admissible to the ontology, there must be clear answers to questions about when we have the same thing and when we have two different things. The predicates of the theory provide a definition of equivalence for propositions, and one of the postulates states that equivalent propositions are identical.

[2] The commitment to sets of propositions is convenient, but it could be avoided. One way to do this is as follows: since the postulates of our theory entail that every set of propositions corresponds to an equivalent proposition, we could have restricted the domain to single propositions but allow for arbitrary mereological fusions, where the fusion of the members of a set of propositions is the proposition equivalent to the set.

Here is a sketch of the simple theory. The domain consists of propositions and sets of them. There are just two primitive monadic predicates, 'C' for *consistency* and 'T' for *truth*, the first applying to both propositions and sets of them, the second applying just to propositions. I will start with some definitions of predicates that are definable in terms the primitive predicates, since the defined predicates will be useful for stating the postulates of the theory. I state both the definitions and postulates in prose, but it will be clear how to represent them in the first-order formalism.

(D1) Two sets Γ_1 and Γ_2 are *equivalent* iff for every set Δ of propositions, $\Gamma_1 \cup \Delta$ is consistent iff $\Gamma_2 \cup \Delta$ is consistent.

(D2) A set Γ *entails* a proposition x iff $\Gamma \cup \{x\}$ is equivalent to Γ.

(D3) Propositions x and y are *contradictories* iff $\{x,y\}$ is inconsistent, but for every consistent set of propositions Γ, either $\Gamma \cup \{x\}$ is consistent, or $\Gamma \cup \{y\}$ is consistent.

(D4) A set of propositions Γ is *maximal consistent* iff it is consistent, and for every proposition x, if $\Gamma \cup \{x\}$ is consistent, then $x \in \Gamma$.

(D5) A proposition x is *necessary* iff x is entailed by every set of propositions.

(D6) A proposition is maximal consistent iff it is equivalent to a maximal consistent set.

The postulates of the theory are as follows:

(P1) Every subset of a consistent set of propositions is consistent.

(P2) The set of all true propositions is maximal consistent.

(P3) Every proposition has a contradictory.

(P4) For every set of propositions Γ, there is a proposition x such that $\{x\}$ is equivalent to Γ.

(P5) For every proposition x, if $\{x\}$ is consistent, it is a subset of a maximal consistent set.

(P6) For any propositions x and y, if $\{x\}$ is equivalent to $\{y\}$, then $x = y$.

These postulates entail that the propositions form an atomic complete Boolean algebra.[3]

The abstract structure of the theory is clear enough, and unproblematic, but the commitment we make is not just to a piece of mathematics. We need to say what the things are to which we are attributing this structure. Following Quine's regimentation procedure, we use the resources of the pre-regimented language—its vocabulary and our understanding of it—to specify (roughly) what we are talking about, and then we examine this notion critically, specifying a regimented term that is not claimed to be synonymous with the pre-regimented word, but is a revision of it. Perhaps it resolves ambiguities that are judged to infect the given term and answers open questions about its application. Our informal characterization of propositions will start with the observation that they are things that are true and false, and so that have truth conditions. (In the end I will say that they *are* truth conditions.) They are the contents of thought and speech—things that are conveyed in communication, and that play an essential role in characterizing how we represent the world.

[3] I gave these postulates in Chapter 2 of Stalnaker 2012 and made this claim about them, but the claim that the postulates are sufficient for a Boolean algebra that was both atomic and complete should have been identified there as a conjecture, since I hadn't proved it. Peter Fritz pointed out to me that it was not obvious, but he also proved, in a more general setting, that it is correct. See Fritz 2021.

2.3. A defense of propositions

What's wrong with propositions, according to Quine? He was certainly right that the word "proposition" and corresponding words in other languages have been used in diverse ways throughout a long theoretical tradition in logic, the philosophy of logic, and the philosophy of language and mind, and he was suspicious that the "philosopher's tolerance toward propositions has been encouraged partly by ambiguity in the term."[4] Propositions are sometimes (particularly in the medieval tradition) identified with declarative sentences themselves, and at other times they are taken to be entities distinct from the sentences that express propositions. But use-mention confusion—Quine's favorite mistake—is not the main problem. A commitment to propositions is a mistake, according to Quine, even if we are clear that we mean to be talking about entities that can serve as sentence meanings, or as what declarative sentences are used to say. Furthermore, his "objection to recognizing propositions does not arise primarily from philosophical parsimony" or "from a disapproval of intangible or abstract objects." The "more urgent" problem, he says, is that if we take propositions to be something like sentence meanings, this will presuppose an equivalence relation between sentences—a notion of synonymy—that "makes no objective sense." Quine takes his arguments against the analytic-synthetic distinction, and for the indeterminacy of translation to be arguments for this negative conclusion.

Quine is here thinking of a commitment to propositions as motivated by a method of characterizing abstract objects that was pioneered by Frege: one begins with an equivalence relation defined on objects that are already ontologically acceptable, and then introduces a kind of entity in terms of that equivalence

[4] Quine 1986, 2.

relation. The concept of number was the most important notion that Frege explained in this way, but a familiar simpler example that he used is the geometric concept of a direction: one paraphrases 'line A is parallel to line B' as 'the direction of line A = the direction of line B'. The identification of these Fregean Thoughts is supposed to justify the introduction of the abstract objects—directions—by giving them clear identity conditions. Quine would agree that if we had a clear notion of sentence synonymy, we could justify an ontology of propositions by following the Fregean pattern, but we don't. I am sympathetic to Quine's rejection of the notions of analyticity and synonymy, but I think it is a mistake to think of propositions as entities that are grounded in semantic relations between sentences, rather than as things that sentences may be used to say, but that are explained and justified independently of the natural language sentences that are used to say them. The theorist can construct regimented languages that aim to establish a tight semantic correspondence between sentences of that language and propositions that they express, but one should think of such languages as presupposing the notion of proposition. Languages in general are, among other things, devices for conveying information, and we will need to tell a long and complex story about how our knowledge of and competence with a language allow us to use its sentences, in context, to convey information. A semantic theory for a language will inevitably have empirical presuppositions (that is what the Quinean manifesto, discussed in Chapter 1, implies), and there may be no clear line between the knowledge that constitutes linguistic competence and the knowledge of matters of fact that tends to be shared and exploited by users of the language in their aim of conveying information. This Quinean observation is good reason to be skeptical of a notion of sentence meaning, but not thereby necessarily a reason to be skeptical of propositions, as pieces of information that language is used to convey.

But if propositions as items of information have the identity conditions that our theory gives them, then (as Quine and many others have noted) they seem to be too coarse-grained to be suitable values for the objects of propositional attitudes.[5] One problem that fuels skepticism about propositions is that they are asked to do too much. The familiar puzzles about belief are connected with considerations that motivate Quine's rejection of the analytic-synthetic distinction: Just as the relation between the meanings of sentences of a natural language and the propositions they express would require a complicated story even if the notion of a proposition were clear, so we should expect that an account of the role played by propositions (with the coarse-grained identity conditions we have given them) in the description of cognitive and conative states of rational agents will be complicated. Kripke famously remarked, when discussing his puzzle about belief (a variation on the familiar Frege puzzle about beliefs ascribed with proper names) that the apparatus of propositions may break down here.[6] But note that what may break down is the assumption that a notion of proposition with identity conditions defined by necessary equivalence can be straightforwardly used to characterize beliefs, and that is not a reason to think that the notion itself is problematic. We can agree that the puzzles about linguistic meaning and propositional attitudes are not dissolved by the introduction of our notion of proposition, but I would argue that those puzzles are sharpened and clarified by distinguishing a notion of proposition that is not directly grounded either in semantic relations between sentences, or in relations between the cognitive roles of sentential clauses.

[5] On the account of propositions I have sketched, sentences of a regimented language that are logically equivalent will express the same proposition, and Quine remarks that "to identify propositions on this basis would disqualify them as objects of belief" (Quine 1960, 204).

[6] Kripke 1980, 21. See also Kripke 1979.

Quine actually showed some sympathy, in *The Philosophy of Logic*, for a notion of proposition as *objective information*.

> The notion of information is indeed clear enough, nowadays, when properly relativized. . . . It makes sense relative to one or another preassigned matrix of alternatives. . . . But a trouble with trying to equate sentences in real life, in respect of the information they convey, is that no matrix of alternatives is given; we do not know what to count. . . . The question when to say that two sentences mean the same proposition is consequently not adequately answered by alluding to sameness of objective information. This only rephrases the problem.[7]

Right, but as Quine would certainly agree, sometimes it is helpful to rephrase a problem. This passage reinforces the judgment that the source of Quine's skepticism about propositions in the assumption that they are being introduced to provide a solution to a problem about linguistic meaning, and we can agree that if propositions were identified with the meanings of sentences of natural language, this would presuppose rather than explain sameness of meaning. Propositions *are* introduced in the hope that they can contribute to an account of the ways that "sentences in real life" are used to convey information, and more generally an account of how they can be useful in the daunting tasks of explaining how language is used to do what we use it to do, and how our beliefs, intentions, and other intentional states relate us to the world that we think about and act in, but there is no presumption that their role in such accounts will be simple or straightforward.

We can agree that no matrix of alternatives is given by the general concept of objective information, but that is what one should expect, given the Quinean picture. It is analogous to his point that

[7] Quine 1986, 3–4.

no predefined ontology is given by the project of developing a regimented theory that presupposes an ontology. On Quine's picture, we construct a regimented language for describing the world, or some aspect of it, while at the same time constructing a theory that says what the world, or that aspect of it, is like. As discussed in Chapter 1, this holistic project does allow for some division of labor. Even if the regimented language we construct inevitably has substantive presuppositions, we can describe an abstract structure that will be common to alternative answers to the substantive questions. For Quine, the project of regimentation involves hypothesizing an ontology and a lexicon for describing that ontology. Two theorists might agree that a theory should take that form even if they disagree about what ontological commitments to make, and about what predicates to admit to the lexicon that is used to make them. Similarly, two people might agree that a project of articulating a substantive theory of propositions should have a certain structure, while disagreeing about the nature of the elements of that structure—the matrix of alternative possibilities. On this way of thinking about propositions, a project of regimentation, in general, might be thought of as a project of constructing a language that *expresses* propositions (about a certain ontology), where propositions themselves might or might not be among the things in the world that that language can talk about.

I argued in Chapter 1 that a Quinean project of regimentation need not aim to be comprehensive—that the ontology, to be realistic, need not be a domain of absolutely everything there is. The account of propositions I will develop makes no claim to be a theory of some domain of absolutely all the propositions there are, and so need not require some absolute most fine-grained matrix of alternative possibilities. But as Quine grants, the notion of proposition as objective information seems reasonably clear, given a context of alternative possibilities. At least it is worth exploring the hypothesis that there are things of this kind that we can theorize about, so let us look further at it.

2.4. Referring to and expressing propositions

Our theory will assume that propositions are among the things we can refer to, quantify over, and say things about, but given the kind of things we are taking propositions to be, it is also reasonable to assume that they are things that are *expressed* by the sentences of the language in which we state our theory. So propositions will be playing two roles in a theory of propositions: they will be things that the sentences of the language say, but they will also be things in the domain of the first-order theory that sentences of the language describe. I will be focusing on the construction of a language and a theory in which propositions play both of these two different roles, but we should expect that the second assumption—that the sentences of the regimented language express propositions—will apply more generally to any such language that we are prepared to endorse. That is, if there are propositions, then it should be a constraint on the acceptability of *any* regimented language/theory that the sentences of the language express propositions. This means that the compositional semantics for a regimented language should tell us what propositions are expressed by the complex sentences as a function of the values of their constituents.

I am here suggesting that a sentence of a regimented language (whatever the domain of entities that constitutes the subject matter of that language) should have a proposition as its semantic value, but I am not thereby taking back the point made in section 2.3 that we are not *presupposing* a notion of sentence meaning or synonymy, and trying to use it to explain the identity conditions for propositions. The order of explanation goes the other way: we defend the intelligibility of a certain notion of proposition with a certain structure, and then take a theory of propositions to be a theory about the kind of thing that language is used to convey. The relationship between the sentences of natural language and the information they aim to convey is complex and mediated by context; our theory of propositions is not an account of that complex relation; it is an account only of one term of it. But the suggestion here is that we might think of a regimented language in general as a representation of some of the propositions that we are capable of expressing.

Because of the closure conditions that our theory imposes on a domain of propositions, the task of giving a compositional semantics that defines the propositions expressed by the complex sentences of a regimented language in terms of the propositions expressed by their parts will be straightforward on the level of sentential logic. First, the theory tells us that every proposition has a unique contradictory, so we can assume that if a sentence ϕ expresses a proposition, then the proposition expressed by $\sim\phi$ will be the contradictory of the one expressed by ϕ. Second, the theory tells us that every set of propositions is equivalent to a unique proposition, so if sentences ϕ and ψ express the propositions x and y, respectively, we can assume that $(\phi \wedge \psi)$ will express the proposition that is equivalent to $\{x,y\}$. The point is that given propositions, and the structure we have assumed that they have, it is straightforward to extend any function from truth-values to truth-values to a function from propositions to propositions, and furthermore, it is also straightforward to add an *intensional* operator[8] if it is justified by the closure conditions on propositions. It is a theorem of our theory of propositions that there is a unique necessary proposition, so the following semantic rule for a necessity operator is well-defined, on the assumption that the sentences of a regimented language express propositions: For any sentence ϕ, let $\Box\phi$ express the necessary proposition if ϕ expresses the necessary proposition, and let $\Box\phi$ express the unique impossible proposition (the contradictory of the necessary proposition) if ϕ expresses a non-necessary proposition. So our assumption that the sentences of a regimented language express propositions justifies extending the sentential logic of such a language to a modal logic.

[8] That is, an operator that is not truth-functional, but that takes propositions to propositions.

Justifying the assumption that the sentences of a regimented language that involve predication and quantification express propositions will be a more difficult matter that we will discuss below and in Chapter 3. For the moment I will just assume that the sentences of a regimented language that are formulated with the resources of first-order quantification theory do express propositions. Or perhaps the point is better put this way: to endorse or accept a certain regimented language is to accept that its sentences express propositions, which requires that one accept that the conditions on admissibility of predicates to the lexicon of the language ensure that they do.

I am suggesting that the logic of the language in which we formulate a theory to clarify our ontological commitments should be a *modal* logic, but Quine emphatically rejected the hypothesis that modality was a part of logic. In our discussion in Chapter 1 of Quine's views about the scope of logic, we postponed a consideration of his reasons for this. Before continuing with the development of the theory of propositions, I will digress to consider those reasons now.

2.5. Grades of modal involvement

Modal logic, Quine argued, was conceived in sin—the sin of confusing use and mention. More specifically, it is a confusion of properties of and relations between *sentences*, on the one hand, and features of the *values* of sentences, on the other.[9] If you look back at the writings of the founder of modern modal logic, C. I. Lewis[10] (who was one of Quine's teachers), you will see that Quine is exactly right about the confusion involved in the origin of the theory (though we must be wary of visiting the sins of the fathers

[9] Quine criticized modal logic in many places, but I focus here on Quine 1953, in which he characterizes the three different grades of commitment to modality.

[10] See, for example, Lewis, C. 1918.

onto their children.) The concepts that Lewis meant to represent in his modal logic were syntactic or proof-theoretic concepts. The strict conditional hook was introduced to express *derivability*, the box and diamond to express *theoremhood* and *proof-theoretic consistency*. Quine had no problem with at least some versions of the first grade of modal involvement—necessity, possibility, and implication as meta-logical predicates, of the sentences of some logistic system, with its axioms and rules of inference. But he argued that the attempt to use these predicates to define operators in the language involved a confusion of expressions with the values of expressions. Where 'N' is the meta-logical necessity predicate (predicating theoremhood or logical validity of sentences of a given logistic system), the necessity *operator* was introduced by stipulating that for any sentence ϕ, $\ulcorner\Box\phi\urcorner$ shall be true if and only if N(ϕ) is true.[11] So for example where '($p \vee q$)' is a sentence of propositional calculus, "N'($p \vee q$)'" says that the sentence '$p \vee q$' is a theorem, or that it is a tautology, which of course is false, whether the system is extensional propositional calculus or the modal language being defined. It might therefore seem that the sentence with the necessity *operator*, '$\Box(p \vee q)$' should be false, and since its falsity is independent of the interpretation of 'p' and 'q', one might expect '$\sim\Box(p \vee q)$' to be a theorem of the modal logic being defined, but no one at the time even considered that it might be. This modal sentence will be true for some *substitution instances* of the sentence, and as Quine often noted, some people confuse values of variables with expressions that may be substituted for those variables. If we

[11] I use Quine's corner quotes here (for the first and last time in this book), since Quine's concerns about use and mention are particularly salient. Corner quotation marks were introduced by Quine to make general claims, using meta-linguistic variables, about complex expressions. 'ϕ' is a variable ranging over sentences of the formal language, and the expression '$\ulcorner\Box\phi\urcorner$' refers to the string of symbols that results from concatenating the logical symbol '$\Box$' with the sentence that is the value of 'ϕ'. For the most part in this book when I am discussing meta-logical issues, I will presuppose less obtrusive conventions, just using unquoted expressions such as 'ϕ' and '$\sim(\phi \wedge \psi)$', for example, to refer to the strings of symbols that result from concatenating the logical symbols with the values of the syntactic variables in the way pictured.

succumb to this temptation and take substitution instances to be values, then we will get the more plausible result that '$\Box(p \vee q)$' is satisfiable, since we can, for example, substitute '$\sim p$' for 'q', yielding a tautology. But as Quine observed in his critique of modal logic, there will be further questions about iterated modality that are not answered by the stipulation that a sentence of the form ϕ shall be true if and only if the sentence substituted for 'ϕ' is a theorem or is valid. For example, one may ask whether $\Box\Box\phi$ should be a logical consequence of $\Box\phi$. We need to extend the necessity predicate to the new sentences with boxes and diamonds in them in order to interpret the box and diamond themselves, and even if the necessity predicate was well defined for the original language, there will be different ways of extending it.

The third grade of modal involvement is the move from an operator on closed sentences (what Quine called "statements") to an operator on both open and closed sentences. C. I. Lewis concerned himself only with propositional modal logic, and so did not make the move to the third grade, but it was a natural move (made later by Rudolf Carnap and Ruth Barcan Marcus), given the operators and connectives introduced by the move to the second grade. Quine's argument in his original critique was that the use-mention mistake was made at the move from the first to the second grade of modal involvement but that the most problematic consequences of this error come out only with the move to the third grade, which involved quantification into something that is equivalent to a quotation.

Now let's compare the way of introducing a modal operator that I suggested at the end of the previous section with C. I. Lewis's way of explaining such an operator. My proposal is like Lewis's move from the first to the second grade of modal involvement in that it uses a *predicate* (the defined necessity predicate of propositions) to give the semantics for an *operator*, but it contrasts with Lewis's move in two ways. First, in our case the predicate at the first grade of modal involvement is a predicate of *propositions*, which are the semantic values of sentences, and not sentences substituted for them. So there is no confusion of sentences with their semantic values. Second, the semantic rule for the necessity operator that I proposed gave the value of the complex sentence containing the operator directly in terms of the value of the sentence that is the operand. It does not, as Lewis's proposal did, equate the meaning of the complex operator sentence with the meaning of the sentence that applies the predicate of necessity, and so it did not leave unanswered the questions about iterated modality. The propositional modal logic that the semantic rule I proposed validates is S5.

In giving this semantic rule, we assumed that the sentences of *any* proper regimented language express propositions, including languages that may not have propositions themselves in their domains. The role of the theory of propositions in stating this rule was in the meta-language in which we give the semantics for the logical operators and connectives, and while the theory of propositions is essential to the *theory* of the modal logic we are introducing, we are not assuming that a language with that logic is necessarily a language that talks *about* propositions. Even though we are using the theory of propositions to explain and justify the introduction of a modal operator to the logic, we are not suggesting that the theory of propositions is itself a part of logic. Compare what we said in Chapter 1, following Quine, about the role of set theory in a theory of logic. The model-theoretic semantics for extensional quantification theory presupposed and used set theory to explain and justify the principles of reasoning that are used in regimented languages, but the reason set theory is still not to be thought of as a part of logic (according to Quine) is that languages based on such a logic need not have sets in their domains, and they need not have the predicates of set theory—in particular the membership relation—in its lexicon. We can formulate set theory in such a language, but we are not talking *about* sets simply by using a language with this logic. I want to say an analogous thing about the relation between modal logic and the theory of propositions. The theory of propositions plays a role in giving and justifying a semantics for the logic, and we can also formulate the theory of propositions in a language with that logic, but a language with that logic *need* not have propositions in its domain, and it need not have the predicates of the theory of propositions in its lexicon.

To help make clear the relation between the necessity predicate that is defined in the regimented theory of propositions and the necessity operator defined in the semantics for the modal logic, let me consider some questions about them that I have been asked. First question: why assume that there is just one necessary proposition? The thesis that there is just one necessary proposition (i.e., just one proposition in the extension of the necessity predicate) is a theorem of the extensional theory of propositions, and so is independent of the proposal to enrich the logic to a modal logic. The question why this should be a theorem of the theory is like the question, why should sets obey a principle of extensionality? Sets are by definition things satisfying that principle. This is not to say that the question whether there are such things as sets can be answered by a stipulation; what is stipulated is that sets, if there are such things, satisfy that principle. Similarly, in our regimented theory we stipulate that propositions should be things that conform to the principle that equivalent propositions are identical, should there be such things. It is recognized that there might be different ways of regimenting the intuitive idea of proposition that we find in our pre-regimented theory, and we don't exclude the possibility that there are alternative ways of regimenting the notion that have finer-grained identity conditions (i.e., we don't exclude the possibility that there might be a legitimate notion of proposition according to which propositions that are equivalent in our sense are not identical). But part of the motivation for including the postulate in our theory of propositions is that other plausible candidates (e.g., perhaps Russellian or Fregean theories of propositions) would satisfy an equivalence relation of the kind our theory defines, and so would allow for an ontological commitment of the kind of entity our theory hypothesizes.

This is the sense in which a theory of propositions satisfying this postulate is minimal.[12]

Second question: Why should the necessity *operator* of the modal logic we are defining be an S5 logic—a logic in which sentences of the form $(\Diamond\Box\phi \rightarrow \Box\phi)$ are valid? Given the assumption that we are making that the sentences of any regimented language we are prepared to endorse *express* propositions, we are in a position to define (in our semantic meta-language) any operator that corresponds to a well-defined function from propositions to propositions, and the semantic rule we have defined satisfies this condition. We know by the principle of excluded middle that every proposition expressed by a sentence is either the unique necessary proposition or it is not, so the semantic rule that $\Box\phi$ shall express the unique necessary proposition if ϕ expresses that proposition, and the unique impossible proposition otherwise is a well-defined rule, and it would be well-defined in the semantic meta-language for any regimented language all of whose sentences express propositions, even if that language did not initially have modal operators. That is why we make no further commitment in adding it to our language. We don't foreclose the possibility that there might be other necessity operators that have a weaker modal logic, though I don't see how to define one without assuming some additional structure. Kripke models for weaker modal logics of course do assume additional structure—a binary accessibility relation between possible worlds.

Third question: In the regimented modal language for our theory of propositions, what *proposition* is expressed by a sentence of the form N*x* for any value of *x*? This is a harder question that points to a more general problem. Our extensional theory of propositions determines the *extension* of the necessity predicate (the *truth-value*

[12] See Stalnaker 2012, 26, for this defense of that postulate. Our assumption that other notions of proposition would allow for an equivalence relation of the kind we have defined is compatible with Quine's thesis, mentioned in section 2.3, that there is no notion of sentence meaning (for sentences of a natural language) that satisfies an equivalence relation of this kind.

of any sentence of the form Nx of the regimented language of the theory of propositions, for any value of x). It implies that Nx is true if and only if x is the unique necessary proposition, but it does not tell us which true proposition is expressed in the case where x is the necessary proposition, or which false proposition is expressed in the case where the value of x is not the necessary proposition. The answers to these questions are not determined by the answer to our second question: they are questions, not about the operator, but about the semantic roles of the predicates in the regimented language of the theory of propositions. I will return to this third question in section 3.4 of the next chapter, where I discuss the lexicon of the regimented theory of propositions, but we must first look at some general questions about modal semantics.

So long as we stay on the level of a sentential logic, our way of moving from the first to the second grade of modal involvement is unproblematic, but the assumption that the sentences of a first-order quantificational language express propositions brings to the surface some further general questions, first about the semantic values of the primitive predicates of a regimented language, and second about variable-binding and the formation of complex predicates. A semantic rule for atomic sentences of a quantified modal language has to say how the proposition expressed by a sentence of the form Fx is determined by the semantic value of the predicate F, and the value of the variable x, and this means we need to say what kind of thing the value of a predicate is. Second, the variable-binding operation is a way of forming complex predicates, and our semantics needs a rule for saying how the value of the complex predicate is a function of the values of the expressions that are its constituents. So even if our move to the second grade of modal involvement did not involve any use-mention mistake, this move, and the further move to the third grade of involvement, still bring new questions that need to be addressed.

A follower of Quine might say that the new questions raised by the assumption that the sentences of a regimented language express propositions just make clearer that it was a mistake to allow them into our ontology in the first place, but I want instead to conclude that the initial ontological commitment brings to the surface questions about the semantics of the first-order language that need to be answered by the Quinean project in any case. The more stringent compositional demands that the enrichment of the language requires focus attention on the constraints on the *lexicon* of the regimented language we are constructing. That is, they focus attention on questions about what is required for a predicate to be *admissible* in the language we are constructing to represent our theoretical commitments. This is a question that Quine, in his construction of a regimented language, addressed only informally. His focus was on constraints on admissibility to one's ontology—the entities in the domain of things that there are to be talked about—and not on the constraints on the predicates we use to talk about those things. We can agree that predicates are not names, in either an extensional or an intensional language, but whether we have propositions in our ontology or not, there have to be some constraints on what a theorist needs to know or presuppose about a predicate in order to admit it into the regimented language that he or she is constructing, a language that is designed for stating in a clear and precise way what the theorist takes the world to be like. These constraints will include commitments about what there is to talk about but also commitments about what can be said about the things we talk about. In fact, we cannot even make sense of an ontological commitment except in terms of the predicates that say what kind of thing our theory is about, and how those predicates distinguish between the things in the ontology.

Quine makes much of the need for identity conditions for the things admitted to one's ontology, but as many philosophers have noted, the notion of identity conditions is hard to pin down. The idea seems to be that when we have different ways to refer to things, there must be answers (presumably provided by one's theory of the objects) to questions about when we are referring to the same thing and when we are referring to different things. We need not necessarily know the answers to these questions—Ralph may not be sure whether the man in a brown hat that he saw in the bar (Ortcutt, as it happens) is the same man as the one he saw at the beach, but it must somehow be implicit in a theory that is committed to the existence of human beings that the facts will determine answers to such questions. In general, one's theoretical commitments are commitments to there being answers to certain questions, and the questions will be stated with the predicates of the theory. So quite aside from an ontological commitment to propositions, the Quinean project needs to say something more systematic about the standards for the admissibility of descriptive predicates to a regimented language if it is to have a clear account of just what ontological commitment involves. The commitment to propositions, I will argue in the next chapter, helps to provide a framework for clarifying and addressing this issue.

3

Predicates and Predication

The hypothesis that we began exploring in Chapter 2 is that our ontology should allow for the existence of propositions. It seemed an innocent further assumption, given the kind of thing that propositions are, that the sentences of a theory in which we talk *about* propositions also *express* propositions, and more generally, if there are propositions, then the sentences of any acceptable regimented language should be expected to express them. But that assumption brought to the surface a cluster of further questions, beginning with some specific questions about the predicates of the theory of propositions and leading to more general questions about the role of the predicates in any regimented language. My task in this chapter is to explore those questions. I will be arguing that the hypothesis that there are propositions meeting the conditions that our theory of propositions imposes provides some resources for addressing them.

3.1. Informal constraints on admissible predicates

Recall that for Quine, a philosophical project involved the specification of two interconnected components: an ontology and a range of predicates for describing it. To endorse a particular regimented language/theory is to make an ontological commitment to its domain and to regard the predicates as admissible. Quine's focus, in characterizing the logical form of his canonical languages, was on ontological commitment: the familiar slogan was that to be is to

Propositions. Robert Stalnaker, Oxford University Press. © Oxford University Press 2023.
DOI: 10.1093/oso/9780197647035.003.0004

be the value of a bound variable. Variable-binding, I have argued, should be regarded as an operation of complex predicate formation, and this suggests that Quine's slogan should be interpreted as saying that to be is to be a subject of predication (or more exactly, to take something to be is to take it to be a subject of predication).[1] The question what predicates are admissible is thus tightly linked to the question of what ontological commitments we are prepared to make.

While Quine was strict in the constraints he imposes on the compositional structure of his regimented languages, and explicit about ontological commitment, he was casual about the assessment of the predicates that should be admissible to their lexicons. For example, he argued that "the subjunctive conditional has no place in an austere canonical notation for science" but then notes that "the ban on it is less restrictive than would at first appear. We remain free to allow ourselves one by one any general terms we like, however subjunctive or dispositional their explanations."[2] But he can't really mean "any general term we like." The "erratic quality" of the examples Quine used to criticize the clarity of subjunctive conditionals in the general case (e.g., If Caesar had been in command in Korea, would he have used the atom bomb or catapults? If Bizet and Verdi had been compatriots, would they have been Italian or French?) would compromise some of the informal

[1] So on the Quinean meta-ontology, something like what has come to be called 'the being constraint' ("to have properties or stand in relations, it is required to be something") is a triviality: it is not a metaphysical thesis, but a statement of the form that a metaphysical thesis takes. (Since Quine did not accept the existence of properties and relations, he would not have put the being constraint this way; his version would say that for a predication sentence of a regimented language to be true, the thing to which the predicate is applied must exist.) It is not only Quineans who see the being constraint as a triviality. Timothy Williamson, who coined the label for this thesis, posed this rhetorical question: "How could a thing be propertied were there no such thing to be propertied? How could one thing be related to another were there no such things to be related?" But others have questioned the being constraint, and we will consider later (in Chapter 5, section 5.3) what it might mean to deny it.

[2] Quine 1960b, 225.

subjunctive explanations we might be tempted to use to introduce a dispositional predicate. The difference, Quine suggested, between legitimate and illegitimate explanations of dispositional predicates is that in the good cases there is a "stabilizing factor." Dispositional predicates are okay if they are "conceived as built-in enduring structural traits," or "insofar they can be fairly viewed as expressing dispositions."[3] Quine was willing, in his informal characterization of standards for the admissibility of predicates, to talk about traits, structural characteristics, and dispositions expressed by a predicate. One might be tempted to say that the process of admitting an empirical predicate to one's lexicon is a process of reference fixing involving the empirical hypothesis that there exists an attribute with certain characteristics. That is, one might be tempted to say that a predicate is admissible only when it expresses a property. In the case of dispositions, on this way of thinking, one hypothesizes that there is a property that tends to manifest itself under certain conditions, but that may be instantiated by an object even when the manifestation conditions are not met. Any such hypothesis would be empirical, and so might be false, but that is just an instance of the general Quinean point that the endorsement of a regimented language brings with it the endorsement of a theory that makes substantive claims about the world.[4]

The one non-negotiable constraint on predicates admissible to a canonical language is that enough information must be presupposed to determine an extension for the predicate, but normally predicates are not introduced by specifying their extensions explicitly by enumeration. A predicate may be unproblematic (from Quine's point of view) even if its extension is unknown or if it is controversial whether some individual is in its extension.

[3] Ibid., 223, 224.

[4] Quine's views about dispositional predicates closely follow those expressed by Nelson Goodman in N. Goodman 1983 (first published in 1955).

And it is intuitively clear (aside from any commitment to propositions) that predicates with the same extension do not always play the same theoretical role in stating one's theory (using here just a pre-theoretic intuitive idea of sameness of theoretical role). Given any predicate (say 'green'), we can always use a maneuver inspired by Nelson Goodman to define a different predicate with the same extension. Take any true factual statement: for example that Barack Obama was born in the United States. Let 'groe' be defined as the predicate that applies to green things if Obama was born in the United States but to things that are not green if he was not. It seems clear that one who translated my 'groe' as your 'green' would be mistranslating, even though it gets the extensions right. (If we were allowed to use counterfactuals, we might make the point by noting that if, contrary to fact, Obama had not been born in the United States, then grass would still have been green, but it would not have been groe.) This is not a case of the kind of indeterminacy of translation that Quine argued for. There would be many behavioral differences between the ways those two predicates are used in our linguistic practice, differences that would justify concluding that a translation that mapped 'groe' to 'green' would be determinately mistaken. The point is that specifying an extension, by enumeration or otherwise, is not sufficient for determining the theoretical role of a predicate.

3.2. Predicates and their compositional role

Can we say something more systematic about the conditions for admissibility that predicates must meet to belong in an acceptable regimented language? One cannot ask for a precise criterion, any more than one can ask for a precise criterion for making an ontological commitment. The world determines whether it contains things of the kind that the theories we construct purport to be talking about, and it also determines what there is to be said about what there is to talk about. The rough idea is that a predicate should be admissible only if the facts determine, for any arbitrary individual in an appropriate domain, whether the predicate is true of it (or more generally, of any arbitrary *n*-tuple of such individuals whether the *n*-ary predicate is true of them). But if we assume that the sentences of an acceptable regimented language express propositions, we can perhaps be more precise about two more specific questions: first, under what conditions do two different predicates have the same theoretical role? Second, are there closure conditions on admissible predicates that we can generalize about? That is, can we say something systematic about what further predicates are definable in terms of predicates that are agreed to be admissible? The hypotheses that there are propositions, and that the sentences that we allow into our canonical language express propositions, does provide a basis for motivating systematic constraints of these two kinds, constraints that will determine something analogous to identity conditions for the semantic role of a predicate. The constraints that these hypotheses motivate derive from the identity conditions for propositions, and they go most of the way toward a rationale for admitting attributes (properties and relations) into our ontology along with propositions. Specifically our understanding of a predicate must include enough information to determine the propositions expressed by sentences in which it occurs. This implies, for example, that if a monadic predicate F is admissible, then for each individual *d* in the domain, there must be a proposition expressed by F*x*, where *x* takes the value *d*. That is, there must be a proposition that is a function of whatever counts as the value of the predicate together with the value, *d*, of the variable (where *d* is in the domain that is the ontology). And this implies that two monadic predicates F and G will have the same semantic role in the language only if for each individual *d* in the domain, F*x* and G*x* express the same proposition for the value *d* of *x*. With the help of our necessity operator, we can state this condition as follows:

$\forall x \Box (Fx \leftrightarrow Gx)$. But this condition will not be sufficient for the two predicates to have the same theoretical role. It must also be true, for predicates F and G to have the same compositional role in determining propositions, that a quantified sentence such as $\exists xFx$ must express the same proposition as $\exists xGx$. This requirement is distinct since it could be that every singular proposition that ascribes F to an individual in the domain is identical to the singular proposition that ascribes G to that individual even if the existential propositions expressed with the two predicates were different. To give an example, suppose F is 'being a daughter of Saul Kripke', and G is 'being a son of Hillary Clinton'. Neither predicate, I will assume, is in fact instantiated, and on widely held metaphysical views, there is no actual entity that could have instantiated either of these predicates, and this implies that for every individual d in the domain, both F*x* and G*x* express the same necessarily false proposition for that value of *x*. But each of the propositions expressed by '$\exists xFx$' and '$\exists xGx$' (that Saul Kripke has a daughter and that Hilary Clinton has a son) is a *contingently* false proposition, and either might have been true while the other was false. So since the predicates 'F' and 'G' are the only two lexical items that distinguish the two existential sentences, these two predicates must be playing different compositional roles in determining the propositions.

For the two predicates to have the same compositional role, it must be not only true, but *necessarily* true that anything that instantiates the one also instantiates the other. That is, to ensure that two predicates have the same compositional role, we must assume the necessitation of the coextension condition: $\Box \forall x(Fx \leftrightarrow Gx)$.[5]

[5] This condition entails the necessary condition stated above, since while the modal logic we are using does not validate the converse Barcan formula ($\Box \forall x\phi \rightarrow \forall x \Box \phi$) in general, in the special case of a biconditional with two predication sentences as constituents, the inference that moves the box inside the quantifier is valid. But the converse inference is not valid; that is, ($\forall x \Box (Fx \leftrightarrow Gx) \rightarrow \Box \forall x(Fx \leftrightarrow Gx)$) is not a logical truth, as illustrated by the example. See Chapter 4 for a specification of the first-order modal logic I am using.

3.3. Defining complex predicates

So our theory provides us with necessary and sufficient conditions for the sameness of the compositional role of two predicates, and the logic also provides us with some closure conditions for the admissibility of predicates. By importing the variable-binding machinery of extensional quantification theory into our modal logic, we are presupposing that any predicate that is definable with this apparatus is admissible. This presupposition was implicit in the example, discussed in the previous section, of the two monadic predicates, 'being a daughter of Saul Kripke' and 'being a son of Hilary Clinton'. We took these two predicates to be admissible since each is definable in terms of a binary relational predicate (*being a parent of*), the monadic predicates, *being male* and *being female*, plus quantifiers and names. In general, it seems reasonable to assume that if a binary relational predicate R is admissible, then so will be certain monadic predicates such as being R-related to a particular individual *d*, being R-related to something, or being R-related to itself. In a particular formalization, any predicate definable in this way might be represented by a primitive predicate of the appropriate kind.

This closure condition for the admissibility of predicates will not be controversial for the extensional language. If we know what we are saying when we say that something is F, or that two things are R-related to each other, then surely we know what we are saying when we say that something is non-F, or that a thing is R-related to some unspecified thing, or R-related to some particular thing, or R-related to itself. But it should be noted that when we enriched the language by adding the modal operator '□', we added to the resources for using the variable-binding device to define complex predicates. Even if the modal operator is well-defined as a *propositional* operator, one could question whether we have reason to think, for example, that if you know what you are saying when you say that something is F, you thereby know what you are saying when you say that something is possibly F. To look back to the different grades of modal involvement discussed in Chapter 2, this is to note that we need to take a closer look at the move from the second to the third grade, which I will now do, beginning with a more general consideration of the mechanism of variable-binding, in the context of the assumption that the sentences of a regimented language express propositions.

The basic variable-binding operation, I have argued, is a device for forming complex predicates, and the device works by implicitly turning sentential operators and connectives into predicate operators and connectives. In the context of the assumption that sentences express propositions, the presupposition is that the closure conditions on propositions will correspond to closure conditions on admissible predicates. An informal argument for this presupposition might go like this: A (monadic) predicate expresses a condition that an individual must satisfy for the predicate to be truly applied to that individual. We understand a predicate to the extent to which we understand what the condition is, which means understanding, for any arbitrary individual of the appropriate kind, what condition is imposed on that individual and its environment by the proposition that the individual satisfies the predicate. This characterization of what it is to understand a predicate is an informal characterization, given in the pre-regimented language in which we are assess our proposals for the construction of a regimented language. This assessment includes the assessment of judgments about what predicates are admissible. It is a characterization that can be made precise only case by case as we decide whether the world provides the resources for describing in a particular way the things in the domain of things that our regimented theory purports to be talking about. But elusive as this condition is, it seems reasonable to assume that *if* (1) the condition is satisfied for some monadic predicate F, so that for any arbitrary individual *d*, there is a determinate answer to the question whether *d* satisfies F in a given possible situation, and *if* (2) we have a well-defined propositional operator **o**, an operator that counts as logical in the sense that it is well defined independently of the particular ontology that our language purports to be talking about, then we can conclude that there will be a determinate answer to the question whether an arbitrary individual *d* would satisfy the condition that is the result of applying the operator to a proposition that says, of *d*, that it satisfies the predicate 'F'. The complex predicate, $\lambda x \mathbf{o} Fx$, is therefore admissible. The argument needs to be generalized to account for the way that complex monadic predicates are defined in terms of relations. Suppose 'R' is an admissible binary relation, so that we can assume that there is a determinate answer to the question whether any pair of individuals $\langle d, d^* \rangle$ each in the domain of some possible situation satisfies the condition imposed by the relation R. Then surely there will be a proposition, for arbitrary individual *d*, that says that *there exists* an (unspecified) individual *d** that is R-related to *d*, so the monadic predicate, '$\lambda x \exists y Ryx$' will be admissible, and similarly for the defined predicate $\lambda x Rxx$. Can we also assume that for some fixed individual e, there will be a determinate answer to the question whether for arbitrary individual d in any counterfactual situation, the pair $\langle d, e \rangle$ satisfies the condition? If so, then there will be a well-defined monadic predicate that says of an arbitrary *d* that $\langle d, e \rangle$ satisfies R. I think the answer to this question is also positive, but we should note the following consequence of this judgment that may seem problematic, a consequence that is illustrated in our example of the monadic predicates, 'is a daughter of SK' and 'is a son of HC'. If the fixed individual (SK or HC) is a contingent existent, then the defined monadic predicate will be admissible only if there is a determinate answer whether an arbitrary individual in some counterfactual situation would be a daughter of SK even when the counterfactual situation is one in which SK did not exist. There would be, in such a counterfactual situation, no *propositions* of the form, '*d* is SK's daughter' that exists there, but that does not matter. What matters is that we can give a determinate answer, in the actual world, to the question whether the predicate 'is a daughter of SK' is true or false of some arbitrary individual in that exists in the counterfactual situation. We can be sure that no one could be a daughter of SK if SK did not exist, and it would be true of everyone that he or she is not a daughter of SK. So the assumption that monadic predicates, defined in terms of admissible relational predicates in this way are admissible seems to be justifiable.

The variable-binding device iterates, and so provides a powerful tool for extending the range of admissible predicates. The compositional structure of a complex predicate may involve quantifiers and abstraction operators interleaved with sentential operators, but the intuitive justifications for the closure conditions iterate as well. So consider, for example, the true proposition that Barbara Bush might not have been George W. Bush's mother ($\lambda x \Diamond \sim \mathrm{M}xw(b)$) and the false proposition that GWB might not have had Barbara Bush for a mother ($\lambda x \Diamond \sim \mathrm{M}bx(w)$). If the intuitive arguments sketched above work, then if M is admissible, and b and w are both names of particular individuals, then $\lambda x \mathrm{M}xw$ and $\lambda x \mathrm{M}bx$ are admissible; if these are, then so are $\lambda x \sim \mathrm{M}xw$ and $\lambda x \sim \mathrm{M}bx$, and if these, then so are $\lambda x \Diamond \sim \mathrm{M}xw$ and $\lambda x \Diamond \sim \mathrm{M}bx$.

These are informal arguments that are based on elusive intuitions. The core idea is that we understand predicates, not just in terms of their extensions—in terms of the things to which they truly apply—but in terms of the kind of things they *would* apply to in a range of possible situations. On this view, predication is an essentially modal notion, paralleling the way in which the notion of a proposition is essentially modal. I don't want to minimize the significance of this potentially controversial assumption about predicates, so let me elaborate: The notion of a proposition is essentially modal since it is essential to a proposition that it is identified, not just by its truth-*value*, but by its truth-*conditions*, which means by the truth-value that it *would* have under arbitrary counterfactual conditions. In the same way, a predicate is essentially modal since it is identified, not by just its extension, but by the extension that it *would* have under arbitrary counterfactual conditions, or more precisely, the extension that it has with respect to an arbitrary counterfactual condition.

On the Quinean picture, the hypothesis that a predicate is admissible may have substantive presuppositions, and for the language of an empirical theory, such presuppositions will be empirical. In the context of the assumption that sentences of our regimented languages express propositions, this means that the presuppositions of admissibility may be *contingent*, so we don't assume that an admissible predicate *would* still be admissible if things were different from the way they in fact are. This observation points to a subtle distinction in our intuitive characterization of the criteria for admissibility, a distinction that is particularly salient in the case of predicates defined in terms of particular individuals, such as our example of the predicate, 'being a daughter of Saul Kripke'. The admissibility of this predicate presupposes the existence of Saul Kripke, so it is reasonable to assume that the hypothesis *that there exists a proposition that Saul Kripke has a daughter* also presupposes that Saul Kripke exists. The presupposition, of course, is true, but the proposition itself can be assessed relative to counterfactual situations in which Saul Kripke does not exist. If Saul Kripke had not existed, he obviously would not have had a daughter, and so the proposition is false, relative to such a situation. The upshot is that we can assess a proposition relative to possible situations in which, if they obtained, the proposition itself would not exist, and an analogous point applies to predicates. We assess the proposition by asking whether the predicate 'is a daughter of Saul Kripke' is true of any individual that exists in that possible situation, and we can make that assessment, even if this predicate would not itself be admissible if the counterfactual scenario were realized. And just as a proposition may be *true*, relative to a possible situation in which it does not exist, so a predicate might truly apply to an individual in a possible situation in which the predicate would not be admissible. Since the predicate 'is a daughter of SK' applies to no one relative to possible situations in which SK does not exist, the predicate 'is *not* a daughter of SK' applies to everyone in such situations.

The predicates that most obviously have empirical presuppositions for their admissibility are those definable in terms of contingent individuals, but the Quinean picture suggests more generally that the predicates of the empirical theories we construct acquire their content from the facts about the world they are used to describe. And as Quine would emphasize, the point applies to the common-sense predicates we begin with ('being a desk', as well as 'being a molecule'), and to the predicates we use to describe our phenomenal experience.

As discussed in Chapter 1, Quine's picture allows for at least one predicate (*identity*) that is a *logical* predicate, one whose admissibility does not require substantive ontological presuppositions. Using the variable-binding device, we can define a monadic logical predicate, *existence*, as follows: $\lambda x \exists y x = y$. This is a predicate that applies to everything in the domain. With the help of sentential operators, we can also define a predicate of nonexistence, $\lambda x \sim \exists y x = y$ (which applies to nothing)[6], and a predicate of *possible* nonexistence, $\lambda x \Diamond \sim \exists y x = y$. While the proposition that everything

[6] In a paper about complex predicates (Stalnaker 1977), I said that while there is a predicate of existence, there is no predicate of nonexistence. This was a careless remark that is false, and that is incompatible with the theory there defended and defended here. The point I meant to make was that negative existential statements (such as 'Vulcan does not exist') are not ascriptions of the predicate of nonexistence to something, but rather denials of an ascription of the predicate of existence. The point was that the scope distinction between $\lambda x \sim \exists y x = y(v)$ and $\sim \lambda x \sim \exists x = y(v)$ is semantically significant, if one allows for empty names. Jeremy Goodman, in a paper that develops an argument against contingentism (Goodman 2016), quotes this unfortunate remark of mine, and he bases part of his argument on it, but while the remark merits criticism, the proper response is to retract it.

Goodman described the paper in which this remark was made as a "locus classicus of negative free logic," and it is right that the semantic significance of this particular example of a scope distinction depends on a certain version of free logic, but the main point of the paper was about predicate formation, not free logic. It is right that the scope distinctions between operators inside and outside of complex predicates are relevant to a defense of negative free logic, but these distinctions are important independently of any account of names. Consider the old joke: A says "Only two people can fit into the front seat of a Volkswagen Beetle," and B responds, "Oh really, who are they?" There is a perfectly good predicate, '*can fit into the front seat of a Volkswagen Beetle*', that many but not all of us satisfy. But that predicate is not involved in A's statement on its intended interpretation.

exists is necessary, and so it is not possible that anything satisfy the predicate of nonexistence, there might (for all the logic says) be things that have the property of *possible* nonexistence. Some philosophers (most notably, Williamson 2013) defend the thesis (*necessitism*) that everything exists necessarily, and nothing could exist except what does exist, but I think it would be a mistake to think that this metaphysical thesis is a *logical* truth. At least it is not in the spirit of a Quinean conception of logic, which aims to separate metaphysical from logical commitments, to build this thesis into modal logic. While Quine rejected modal logic, as logic, his arguments for that conclusion were based on the fact that a use-mention confusion was involved in the explanation and motivation provided by the original founder of modern modal logic. I have argued that modal logic can be given a better foundation, based on the hypothesis that the sentences of a regimented language express propositions, and that on this understanding, modal logic can meet the spirit of Quine's general conception of logic—that it can be metaphysically neutral.

Before moving on from the issues about the definability of complex predicates, let me conclude this section with a brief digression about the role of names. I mentioned names as well as predicates among the resources available for defining complex predicates, and assumed, in the examples given, that we might have a name such as 'Saul Kripke' in a regimented language, and that it was a way of identifying a particular individual. For reasons discussed in Chapter 1, Quine banned names from his regimented language, replacing them with predicates that were stipulated to apply to a particular individual. The admissibility of predicates of this kind required the same presupposition—that one has identified a particular individual. Quine argued that this maneuver allowed names to be banned without a loss of expressive power, and I agreed that this was correct—that the maneuver works, even though I preferred a different way of addressing the problems that are raised by having names in the regimented language. The point I want to make now is that the issues about the admissibility of complex predicates are independent of the issues about how to treat names in a regimented language. We could define the same complex predicates (or predicates with the same semantic roles) with Quine's maneuver as we can by having the corresponding names in the language. So if 'S' is a primitive predicate stipulated to apply uniquely to Saul Kripke, and 'D' is the binary *daughter of* relation, then we can define the complex predicate, 'being a daughter of Saul Kripke' as $\lambda x \exists y(Sy \wedge Dxy)$.

Recall that the problem that led Quine to ban names was that if our regimented language had a syntactic category of primitive expressions (names) such that a substantive ontological presupposition was required in order for the name to be an admissible term of that category, then we spoil the division of labor, the ontological neutrality, that the delimiting of the domain of logic was aiming to achieve. The alternative response to this problem that I prefer is to drop the ontological commitment implicit in the classical formulation of a predicate logic with names, allowing for primitive singular terms that fail to refer. As Quine himself observed, it is required, for a name to play the compositional role that it plays, that it have *at most* one individual in its extension, but one could have a syntactic category of expressions that met *that* condition without building specific ontological presuppositions into the logic, and in this way preserve the distinction between the lexicon (which could be specific to a substantive theory) and the resources of the logic (which remain ontologically noncommittal).

There are two ways to understand the category of predicates that Quine used to simulate, or replace, singular terms: on a narrow view, they are predicates that are stipulated to apply to a particular individual that has been identified in the pre-regimented language. On this view, the predicate will be inadmissible if the presupposition that a particular individual has been identified is false. The admissibility of the predicate will thus presuppose the existence of an individual that is its extension, but the presupposition will not be reflected in the logic since the logic does not distinguish between predicates explained in this way and monadic predicates in general.

On a broader view, the predicates that replace singular terms might be explained by a definite description (e.g., "the planet that plays such and such a role in explaining the perturbations of the planet Mercury"). On this view, the predicate will be stipulated to apply to at most one thing, but it might be admissible even if its extension were empty. (The admissibility of the descriptive predicate I just used as an example presupposes the existence of Mercury, but not of a planet that plays a certain role in explaining Mercury's perturbations.) I will assume, in the free logic that I argued is essentially equivalent to the Quinean maneuver, that primitive singular terms may have semantic values that are of the broader kind, so that they may be admissible even if they have no referent. When we enrich the language by adding a modal operator, this distinction will make a difference even in cases where the singular term has a referent. Terms introduced in the narrower way will all be terms stipulated to apply to a particular individual, and so will be rigid designators. Some terms explained in the more general way may be non-rigid designators. The version of free logic I will use does not distinguish the singular terms of the two different types (there won't be different syntactic categories for them), but we can express, in the object language, the condition satisfied by the singular terms of the narrower kind. (See Chapter 4 for details.)

3.4. The lexicon of the theory of propositions

My main point in this chapter has been that the theory of propositions, together with the hypothesis that the sentences of a regimented language express propositions, allows us to ask some more precise questions about the status of the predicates in a regimented language and provides some resources for answering those questions. We have given necessary and sufficient condition for two predicates to play the same compositional role, and we have defended some closure conditions for the admissibility of predicates. This does not give us a precise criterion for the admissibility of a predicate. That should not be expected since decisions about admissibility to the lexicon, like decisions about what ontological commitments to make, are judgments about what the world is like, which are judgments about what overall theories to accept. But the criteria for sameness of compositional role do give us a basis for giving clear identity condition for attributes, should we choose to assume that there are entities (attributes) that are expressed by predicates. We could make this choice with our eyes open (as Quine generously conceded) so long as we recognize that predication is different from naming, and that we are not explaining predication by identifying an entity that is the semantic value of a predicate. I will consider in Chapter 5 the hypothesis that there are entities—properties and relations—that are the semantic values of predicates, but for now let's assume only that admissible predicates have a semantic role, in accordance with the compositional rules, in determining propositions. I will conclude this chapter by looking at the descriptive predicates in the lexicon of the theory of propositions, and at some questions about the semantic roles of these specific predicates.

While the ontological commitment to propositions remains controversial, the theory of propositions, with the predicates it uses to say what such things would be like, seems intuitively clear enough to give us a sense of what is involved in a commitment to propositions. That is, the informal constraints on the admissibility of predicates that Quine uses seem sufficient to justify the admissibility of the predicates we have used to state the theory. The questions about the lexicon of our theory are internal to the theory, but that is what one should expect, given the general Quinean semantic and epistemological holism.

The proposed general criterion for sameness of compositional role was this: Two predicates, F and G, are equivalent if and only if they are necessarily coextensive. Our theory has just two primitive predicates, but it gives us the resources to define predicates that are coextensive with these primitive predicates, but that leave open the question whether they are *necessarily* coextensive, and so whether they have the same compositional role. One case is this: The theory entails that there is a unique necessary proposition and a unique impossible proposition. We might give them names, say 'ν' (nu) for the necessary proposition and 'κ' (kappa) for the impossible (or inconsistent) proposition. (Or alternatively, if you follow Quine in banning names from the regimented language, we could introduce primitive predicates that, like Quine's 'Socratizes', are stipulated to apply to a particular thing that has been identified—predicates that would be stipulated to have the same compositional roles as $\lambda xx = \nu$' and '$\lambda xx = \kappa$'.)

One of the two primitive predicates of the theory is *consistency*, and a proposition is consistent if and only if it is not inconsistent (impossible), so the extension of the primitive predicate C is the same as the extension of the predicate '$\lambda xx \neq \kappa$'. But are these two predicates *necessarily* coextensive? This is the third question raised in the discussion of grades of modal involvement in section 2.5 of Chapter 2, a question we there postponed. The extensional theory of propositions does not provide an answer to this question, even if we assume, as we will, that the postulates of the theory of propositions are all themselves necessary truths. (This assumption is reasonable, since the theory of propositions is a theory about the essential structure of propositions, and not just about a structure they happen to have.) The theory does imply that there must be unique necessary and impossible propositions, and that they must be contradictories, but it does not decide the question whether the *actual* unique necessary and impossible propositions are necessarily the propositions to which those predicates apply. In this case, it seems reasonable to assume that they are, but we should look more closely at just what it means to make this assumption.

There is a parallel question about the second primitive predicate of the theory, *truth*. The theory implies that there is a unique proposition that is both maximal consistent and true, and we might name it 'α'. The predicate of being entailed by α $\lambda x\text{ENT}(\alpha, x)$, where 'ENT' is the binary entailment predicate holding between propositions), is coextensive with the truth predicate, but in this case it is clear that this defined predicate has a different compositional role from the predicate of truth. Suppose *d* is a contingently false proposition, so that the predicate '$\lambda x \Diamond Tx$' is satisfied by *d*. The predicate $\lambda x \Diamond ENT(\alpha, x)$' will, however, not be satisfied by *d*, since *d* is in fact false, and so not entailed, or possibly entailed, by the maximal proposition that is *in fact* true. This implies that '$\Box \forall y(Ty \leftrightarrow \lambda x ENT(\alpha, x)y)$' will have counterexamples if there are contingently false propositions. Even though the reference of 'α' is fixed by definite description ('the unique maximal consistent true proposition'), the proposition that α satisfies this description is contingent.[7]

So the predicate 'is maximal consistent and true' will not have the same compositional role as the predicate 'is the proposition that is in fact maximal consistent and true', even though it is a consequence of our theory that these two predicates will have the same extension. The generalized material equivalence, $\forall y(Ty \leftrightarrow \lambda x ENT(\alpha, x)y)$, will be what Kripke called a contingent a priori truth.

[7] An analogy: compare the proposition that Julius is Julius with the proposition that Julius invented the zip (where 'Julius' is introduced by a reference-fixing stipulation that 'Julius' shall be a rigid proper name for the inventor of the zip). That Julius invented the zip if anyone did is said to be a contingent a priori truth. This is Gareth Evans's story, told with Kripkean jargon, but here is a quasi-Quinean variation: Names are to be banned from the regimented language, but we can achieve the expressive power that names provide in the following way: Suppose we can identify a particular individual that is in an ontology that we are prepared to endorse. We can then introduce a predicate in the regimented language that is stipulated to apply uniquely to that individual. The predicate will be admissible, assuming that we were correct in our assumption that we had succeeded in identifying an individual. This method of introducing a predicate to the lexicon will work even if our means of identifying the individual was not with a name, but with a contingent description. Suppose we say that J shall be a predicate that applies uniquely to the person who is in fact the inventor of the zip. If no one person invented the zip, then the presupposition of our stipulation (that we have succeeded in identifying a unique individual) is false, so the predicate is not admissible after all, but if the presupposition is true, then the predicate is well defined. Now what is the relation between the proposition expressed by 'Jx', for a given value of *x*, and the proposition that *x* invented the zip (for the same value of *x*)? If, contrary to fact, someone else had invented the zip, then '*x* invented the zip' would be true where *x* takes that person as value, but 'Jx' would be false for that value. Furthermore if, contrary to fact, Julius had not existed at all, then the proposition expressed by '$\forall x{\sim}Jx$' would have been true, even if someone else was, in that counterfactual situation, the inventor of the zip. (See Evans 1979.)

Now let's look again at the other primitive predicate of our theory of propositions, and at the relation between the predicate 'is inconsistent' and the predicate 'is a proposition that is in fact inconsistent' (or equivalently, at the relation between the necessity predicate and the predicate of being the unique proposition that is in fact necessary). What is the effect of the assumption that these two predicates have the same compositional role? Here is the intuitive picture: Propositions are ways of distinguishing between a range of alternative maximal possibilities ("a pre-assigned matrix of alternatives," as Quine put it). We are distinguishing between those possibilities from the perspective of the actual world, which is represented by one of those alternatives. We are (again echoing Quine's naturalistic picture) "working from within" rather than from some "transcendental vantage point," using the resources of the world we find ourselves in to distinguish between the alternatives. If the world had been different from the way it in fact is, we might have had different resources for dividing up logical space into matrices of alternatives, but the hypothesis we are endorsing if we say the predicate 'is necessary' has the same compositional role as the predicate 'is the proposition that is actually necessary' is that the different ways we might have of dividing up the space are different ways of partitioning the same space. The necessary proposition is the totality of the space—*the* proposition that would be realized no matter what.

Recall that C. I. Lewis, in his early development of modal logic, explained the necessity *operator* in terms of a necessity *predicate*, and that since his necessity predicate was a predicate of sentences, rather than of the values of sentences, this move involved a use-mention confusion. Our alternative explanation of the necessity operator also used a necessity predicate (this time of propositions) to explain it, but since propositions are the values of sentences, there was no use-mention confusion. Furthermore, while it was assumed that the sentences of a regimented language *express* proposition, the explanation of the operator did not assume, in general, that propositions were in the domain of a regimented language; it assumed only that the sentences of such a language would express propositions. The semantic rule, stated in the semantic meta-language for modal logic, explained the operator directly as a function from propositions to propositions, without making any assumptions about what those propositions were about. So the operator was a well-defined operator for any regimented language for which the sentences express propositions, whatever kind of thing is in the domain of that language. But in the more specific context of the object language for the theory of propositions, which has a necessity *predicate* in its lexicon, a natural hypothesis is that $\Box\phi$ will express the proposition expressed by 'Nx', where x has as its value the proposition expressed by ϕ. It is a natural hypothesis, but whether it is true will depend on how we explain the compositional role of the necessity predicate. If we close the open question raised above in the way suggested, then the natural hypothesis can be shown to be correct, though not by stipulating that it is correct, but by specifying the compositional role of the primitive predicate in terms of which semantic role of the necessity operator is defined. We can close this open question by adding the following modal postulate to our first-order theory of propositions: $\exists x\Box Nx$. This implies that the necessity and consistency predicates have the same compositional role as the predicates defined in terms of the specific propositions, ν and κ.

In the development and defense of an ontology of propositions, there is a subtle interplay between the semantic meta-language for modal logic and the regimented theory that is about propositions, an interplay that illustrates the kind of bootstrap procedure that characterizes the Quinean methodology. We started with a first-order, purely extensional theory of propositions that helped to get clearer about just what kind of thing we purport to be talking about when we hypothesize that there are propositions. But it was then suggested that if an ontology of propositions is defensible, it also seems reasonable to suppose more generally that the sentences of a regimented theory should be taken to express propositions, and this required an enrichment of the logic of regimented languages so that their compositional semantics will determine the propositions expressed when we are talking about things of any kind. The form of the enriched logic was guided by the assumptions made in the first-order extensional theory of propositions about their structure, but it also raised further general questions about the compositional role of the predicates of any regimented theory, questions we argued were good questions about the general project of regimentation, and questions that the theory of propositions helped to sharpen. But then the further questions can be asked about the predicates of the regimented theory of propositions itself, and answering the questions about the specific predicates of the lexicon of that theory motivated an addition to the postulates of the theory of propositions, a further clarification of the kind of thing we purport to be talking about when we talk about propositions.

So we have specified a first-order theory of propositions where the sentences of the theory *express* propositions that are *about* propositions, where the propositions that are in the domain of the theory include the propositions expressed by the sentences of that very theory (as well as propositions that may be about other things). Among the predicates in the lexicon of the theory is a truth predicate, so we have a formal theory with a truth predicate in a language that applies to the values of its own sentences. Are we playing with fire? Painful experience in the history of the development of logic and formal semantics gives us reason to worry about the possibility of paradox in a theory with features like this. Can we be sure that our theory is consistent? There cannot be a *semantic* paradox in the usual sense, since our theory of propositions is not a semantic theory. That is, it is not a theory about the sentences of its language, and the relation between those sentences and their values. But it is a theory about the values of its sentences and about which ones of them are true. We can, however, show that the theory is consistent by showing that any consistent set of sentence of the modal language has a model that can be extended to a model for the theory of propositions. I sketch a construction of this kind in the next chapter.

4

First-Order Modal Logic and a First-Order Theory of Propositions

In Chapter 2, I argued that we should accept an ontological commitment to proposition, along with the assumption that the sentences of a regimented language *express* proposition. The second of these assumptions motivated the decision to take the basic logic of a regimented language to be a modal logic, since it requires the compositional semantics for the language to yield propositions as values for the sentences. Then in Chapter 3, I argued that the move to a modal logic helped to sharpen some general questions about the admissibility of predicates to a regimented language, as well as some more specific questions about the predicates of the theory of propositions. Chapter 3 concluded by raising some worries about the consistency of a theory that is about the propositions that it expresses, and that contains a truth predicate for those propositions. In this more technical chapter, I have two aims: The first is to spell out the details of the modal logic, its syntax, proof theory, and model theory, with a completeness theorem for the logic. The second is to look at the representation of the theory of propositions in this model theory. In that context I will argue that we can have a consistent first-order theory that talks explicitly about a domain of propositions that includes the propositions that are expressed by the sentences of the language of the theory.

The first task is to spell out the particular version of first-order modal logic that I will be using. I will begin by specifying two separate logics—a first-order *extensional* predicate logic with identity,

Propositions. Robert Stalnaker, Oxford University Press. © Oxford University Press 2023.
DOI: 10.1093/oso/9780197647035.003.0005

and a *propositional* modal logic—and then putting them together to get a quantified modal logic. The semantic rules for the extensional logic all generalize nicely to the richer modal language, and the modal model theory will validate the theorems that result from combining the axiom systems for the two logics. Just one axiom concerning the interaction of modality and quantification needs to be added to get a logic that is both sound and complete. The completeness theorem uses familiar methods; I give the details in an appendix.

In the model theory for our first-order modal logic, propositions are modeled by sets of possible worlds, and these will be the values of the sentences—things that the sentences of the language *express*. In the particular model that is defined to prove completeness, there is a denumerable number of possible worlds, and we define a set of sentences for each world that is designed to be the set of all sentences that are true in that world of the model. It will be a corollary of the theorem that for any consistent set of sentences of the language, there exists a model meeting certain conditions, and I will build on that result to give an interpretation for the theory of propositions—a theory in which the propositions expressed by the sentences of the theory are among the elements of the domains over which the first-order quantifiers of the language range. But first I will spell out the details of both a proof theory and a model theory for our first-order modal logic.

4.1. Extensional quantification theory

The extensional first-order logic I will use differs from the more traditional formulation in two ways: first, the variable-binding operator is not a quantifier, but an operator for defining complex predicates; second, the logic is a free logic, allowing non-referring singular terms, and the empty domain. But it will be a bivalent language: predication sentences containing empty names will be false. Despite these differences, the semantics of the variable-binding device is essentially the same as in the traditional Tarski semantics.

4.1.1. Syntax

The *vocabulary* consists of three kinds of primitive descriptive expressions: sentence letters, n-place predicates for each $n > 0$, and individual constants, or names (a denumerable set of each), individual variables (an ordered denumerable set), four logical symbols, '$\forall$', '$\wedge$', '$\sim$', and 'λ', a two-place logical predicate, '=', and parentheses for punctuation. Non-primitive logical symbols, '$\vee$', '$\rightarrow$', '$\leftrightarrow$', are to be understood as abbreviations in the usual way. A *singular term* is either a variable or an individual constant.[1]

The *formation rules* provide a recursive definition of the complex expressions, which are of two kinds: sentences and one-place predicates.

If F is an n-place predicate and $t_1 \ldots t_n$ are singular terms, then $Ft_1 \ldots t_n$ is a sentence.
If ϕ is a sentence, then $\sim\phi$ is a sentence.
If ϕ and ψ are sentences, then $(\phi \wedge \psi)$ is a sentence.
If F is a one-place predicate, then $\forall$F is a sentence.
If ϕ is a sentence and x is a variable, then $\lambda x\phi$ is a one-place predicate.

Since our quantifier is not a variable-binding operator, the formation rule for '$\forall$' is different from the usual one. The one-place

[1] In describing the vocabulary of the language, I have specified the logical symbols, but not the specific descriptive symbols or individual variable. In stating formation rules and axiom schemata, and more generally in talking about the expressions of the language, I use syntactic meta-variables ranging over expressions of the different kinds (ϕ and ψ for sentences, or sometimes for well-formed expressions generally, x, y, and z for individual variables, F and G for predicates, s and t for singular terms, etc.) to represent the form of a sentence. So, for example, the metalinguistic expressions '$\sim(\phi \wedge \psi)$' represents the form of any string of symbols that begins with '$\sim$' followed by '(' followed by a sentence followed by '$\wedge$' followed by a sentence followed by ')'.

predicates that it operates on will normally be complex predicates of the form $\lambda x\phi$, and the sentence that says that everything satisfies $\lambda x\phi$ will be $\forall\lambda x\phi$. Since this will mean exactly the same thing as $\forall x\phi$, in the more familiar formulation, we will normally drop the lambda, taking $\forall x\phi$ to be an abbreviation for $\forall\lambda x\phi$. In effect, this is to treat it as a defined variable-binding operator. The non-primitive existential quantifier will be treated in a parallel way: $\exists x\phi$ will be an abbreviation for $\sim\forall\lambda x\sim\phi$. We can define a predicate of existence, E, as follows:

$$E =_{df} \lambda x \exists y x = y.$$ [2]

An occurrence of a variable x in an expression is *bound* if it occurs within a complex predicate of the form $\lambda x\phi$. An occurrence of a term is *free* if it is not a bound variable. An *open expression* is an expression that contains at least one free variable. An expression that is not open is closed. The metalinguistic expression '$\phi^{s/t}$' denotes the result of substituting s for all free occurrences of t in ϕ (re-lettering bound variables in ϕ if necessary to ensure that all new occurrences of s in $\phi^{s/t}$ are free). (s and t can be either individual variables or individual constants, and a constant is by definition free.)

4.1.2. Proof theory

Axioms: Any closed instances of the following schemata are axioms.

Propositional logic: all (closed) tautologous sentences

Abstraction: $\vdash\forall x(\lambda y\phi x \leftrightarrow \phi^{x/y})$

[2] Strictly, for this to be an unambiguous definition, we need to specify which variables occur in the defined formulas, so let's stipulate that x shall be the first variable in the ordered sequence, and y shall be the second.

Quantification: $\vdash \forall x(\phi \rightarrow \psi) \rightarrow (\forall x\phi \rightarrow \forall x\psi)$

$$\vdash \forall x\mathrm{F}x \leftrightarrow \forall \mathrm{F}$$

Existence: $\vdash \mathrm{F}^n t_1 \ldots t_n \rightarrow \exists xx = t_i$

Identity: $\vdash t_i = s \rightarrow (\mathrm{F}^n t_1 \cdots t_i \cdots t_n \rightarrow \mathrm{F}^n t_1 \cdots s \cdots t_n)$

Rules:

Modus ponens (MP): If $\vdash (\phi \rightarrow \psi)$ and $\vdash \phi$, then $\vdash \psi$

Universal Generalization (UG): If $\vdash (\phi \rightarrow \psi)$ and there are no occurrences of t in ϕ, then $\vdash \phi \rightarrow \forall x\psi^{x/t}$

A *proof* is a sequence of closed sentences each of which is either an axiom or the result of applying one of the rules to sentences that occur earlier in the sequence.

Some remarks on the axiom schemata and rules:

(1) Note that the axioms are all closed sentences, and proofs consist of sequences of closed sentences. So in particular, in a step of a proof applying the rule of universal generalization, one will first prove a closed sentence with an arbitrary individual constant t, and then replace the t with a variable before prefixing the universal quantifier.

(2) Note that the axioms for the quantifier do not include a principle of universal instantiation (UI). This is because the appropriate UI principles are theorems, provable with the help of the axiom of identity. Since the logic is a free logic, the UI theorem for predication sentences, $\vdash \forall \mathrm{F} \rightarrow (\exists xx = t \rightarrow \mathrm{F}t)$, has an existence condition. A second UI theorem is the following: $\vdash \forall x(\forall y\phi \rightarrow \phi^{x/y})$. Sketches of the proofs of these theorems, among others

that are used in the completeness proof, are given in Appendix I of this chapter.

(3) As noted above, our version of extensional quantification theory with identity differs from the traditional formulation, first by being a free logic that allows for the possibility both of the empty domain and non-referring singular terms, and second by having lambda abstraction as the basic variable-binding operator. The provability of the UI principles depends on the first of these features, but not the second. While this has not to my knowledge been recognized in the free logic literature, when identity theory is added to the standard axiomatization of free logic, the UI axiom becomes redundant. (It is not redundant in classical logic, since one needs that axiom to prove $\exists x x = t$, which is valid in a classical system that assumes that all individual constants refer. But if one formulated the classical system simply by adding the axiom $\exists x x = t$ to a version of free logic with identity, the UI axiom would be redundant.

(4) It is normally taken for granted that any consistent relabeling of bound variables in a sentence will result in a provably equivalent sentence (that is, that $\vdash \forall x\phi \leftrightarrow \forall y\phi^{y}/_{x}$ is a theorem), since it is easy to prove this in a standard axiomatization of extensional predicate logic, which has an axiom of UI. As noted, our system lacks such an axiom, but since the usual UI axiom is a theorem of our system, we can also prove the relabeling theorem. Again, see Appendix I for the details.[3]

(5) The existence axiom is not included in standard formulations of free logic, and it is not included in Kripke's

[3] Peter Fritz (personal communication) observed that this could not be taken for granted.

formulation of modal predicate logic. He considers but rejects this principle, giving the following reason in a footnote:

> It is natural to assume that an *atomic* predicate should be *false* in a world H of all those individuals not existing in that world; that is, that the extension of a predicate letter must consist of actually existing individuals.... We have chosen not to do this because the rule of substitution would no longer hold when the atomic formulae are replaced by arbitrary formulae.[4]

This is a bad reason, based on a mistaken conception of substitution. A strict rule of substitution allows for the substitution of complex expressions of some type for simple expressions of that type. But because there are no complex predicates in the traditional formulation of quantification theory, open sentences are treated as if they were complex predicates, and so substitutable for simple predicates. For example, one allows a sentence such as (b) $\forall z\Box(\sim Gz \rightarrow \exists y \sim Gy)$ to count as a substitution instance of (a) $\forall z\Box(Fz \rightarrow \exists yFy)$. If Kripke were to make the natural assumption he suggests, then (a) would be valid, while (b) would not be. But ' $\sim G$ ' is not a predicate, so (b) is not really the result of a substitution in (a). The traditional formulation of predicate logic, without lambda abstraction, builds an equivocation into the notation by treating open sentences as if they were complex predicates. Specifically, the notation implicitly assumes that a sentence of the form (c) $\lambda x \sim Gx(t)$ is equivalent to (d) $\sim \lambda xGx(t)$. A proper substitution rule will allow substituting $\lambda x \sim Gx$ for F in Ft, but not substituting either $\sim G$ or $\sim \lambda xGx$, since these are not predicates, or even well-formed expressions. In the classical semantics for

[4] Kripke 1963, footnote 11.

extensional predicate logic, which assumes that all names refer, the scope distinction between (c) and (d) is not semantically significant, but this assumption should not be built into the notation. And even if it is assumed that all names refer, this will not imply, when we generalize to quantified modal logic, that saying or supposing of x that it is *necessarily* non-G is the same as saying or supposing that it is necessarily not the case that x is G. That is, $\forall x(\Box\lambda y \sim Gy(x) \leftrightarrow \Box\sim \lambda yGy(x))$ will not be valid.

(6) The existence axiom ensures that our free logic is a *negative* free logic. The standard minimal free logic is neutral on the status of predication sentences with non-referring terms: they may be false, lack a truth-value, or be sometimes true and sometimes false. To allow for the last of these alternatives, one would need some additional structure (perhaps a Meinongian outer domain of nonexistent things) to distinguish cases where Ft is true from cases where it is false when t fails to refer to an existing thing. But extensional negative free logic needs no such structure. As we noted above, one can define an existence predicate as follows: $E =_{df} \lambda x\exists yx = y$, but the existence axiom is $\vdash F^n t_1 \ldots t_n \rightarrow \exists xx = t_i$, rather than $\vdash F^n t_1 \ldots t_n \rightarrow \exists t_i$. The axiom must take this form since, while $\lambda x\exists yx = y(t)$ is provably equivalent to $\exists xx = t$, one needs the existence axiom to show this.

(7) There is only one axiom for identity. It can be proved from this that everything is identical to itself, but because this is a negative free logic, instances of the schema $t = t$ are not theorems, and are not valid since an instance of the schema $t = t$ will be false in a model in which t is a non-referring individual constant. But both $(Et \rightarrow t = t)$ and $\forall xx = x$ are theorems. They are included in the list of those proved in the appendix.

4.1.3. Semantics

4.1.3.1. Models and assignment functions

A model for the extensional free logic is a pair $\langle D, [\![\,]\!] \rangle$, where D is a set, the domain of individuals, and $[\![\,]\!]$ is a function assigning values—extensions—to the primitive descriptive expressions of the language. The domain may be empty, and the valuation function may be a partial function, assigning no values to some individual constants (though it must be fully defined for all sentence letters and predicates).

The extension of a sentence is a truth-value: 1 or 0; the extension of an n-place predicate is a set of n-tuples of members of D; the extension of an individual constant, if it has one, is a member of D.

Well-formed expressions in general, including both open and closed sentences and predicates, receive values in a model relative to an assignment of values to the variables. The closed sentences and predicates receive the same value for all assignments of values to the variables. An assignment function, s, is a function from the set of individual variables into the domain. (If the domain is empty, there are no assignment functions.)

If s is an assignment function, then $s[d/x]$ is the assignment function that is everywhere like s, except that for x it takes the value d.

The identity sign, '=' is a two-place logical predicate whose value, for all valuation functions, is given as follows: $[\![=]\!] = \{\langle d, d \rangle : d \in D\}$. Since identity is a two-place predicate, uniform formation rules for atomic sentences would require that identity sentences have the form '$= t_1\, t_2$' but, bowing to tradition, we are using the usual notation: $t_1 = t_2$, and $t_1 \neq t_2$ for $\sim t_1 = t_2$.

Semantic rules:

The semantic rules extend the valuation function to give values, relative to an assignment function, to the complex expressions, which are all either one-place predicates or sentences. If $[\![\,]\!]$ is the valuation function and s an assignment function, then $[\![\,]\!]^s$ is the function that assigns values relative to that assignment function. If x is a variable, $[\![x]\!]^s = s(x)$. If ϕ is a primitive descriptive expression, then $[\![\phi]\!]^s = [\![\phi]\!]$.

The rules for complex expressions are as follows:

Sentences:

If ϕ has the form $Ft_1 \ldots t_n$, then $v[\![\phi]\!]^s = 1$ iff

$\langle [\![t_1]\!]^s, \ldots [\![t_n]\!]^s \rangle \in [\![F]\!]^s$. $[\![\phi]\!]^s = 0$ otherwise.

If ϕ has the form $\sim\psi$, then $[\![\phi]\!]^s = 1 - [\![\psi]\!]^s$

If ϕ has the form $(\psi_1 \wedge \psi_2)$, then $[\![\phi]\!]^s = [\![\psi_1]\!]^s \times [\![\psi_2]\!]^s$

If ϕ has the form $\forall F$, then $[\![\phi]\!]^s = 1$ if $[\![F]\!]^s = D$. $[\![\phi]\!]^s = 0$ otherwise.

Predicates:

If F has the form $\lambda x \psi$, then $[\![F]\!]^s = \{d \in D : [\![\psi]\!]^{s[d/x]} = 1\}$.

The system defined in section 4.1.2 is sound and complete with respect to this semantics. That is, all of the theorems are valid, and every consistent set of sentences is satisfiable in a model, and so every valid sentence is a theorem. This can be proved using familiar methods, with the help of the theorems proved in Appendix I of this chapter. (Note that eleven of the thirteen theorems proved there use only the axioms of the extensional theory.) Since one can see a model for the extensional theory as a special case of a model for the more general modal predicate logic that I will describe just below—a model with just one possible world—the completeness for the extensional theory with be a corollary of the completeness theorem for the modal predicate logic that we prove in Appendix II, though of course a direct proof would be much simpler.

4.2. The modal generalization

We will combine our extensional quantification theory with a propositional modal logic, which is defined as follows:

4.2.1. Syntax

The vocabulary of the propositional modal logic consists of sentence letters, the operators, '~' and '∧', with the same formation rules as in the extensional quantification theory, and one additional symbol, '□', with this formation rule: If ϕ is a sentence, then $\Box\phi$ is a sentence. The possibility operator $\Diamond$ is defined in the usual way: $\sim\Box\sim$.

4.2.2. Proof theory

The propositional modal logic, **S5**, has just three axiom schemata in addition to the tautologous sentences:

$$\vdash \Box(\phi \rightarrow \psi) \rightarrow (\Box\phi \rightarrow \Box\psi)$$

$$\vdash \Box\phi \rightarrow \phi$$

$$\vdash \Diamond\phi \rightarrow \Box\Diamond\phi$$

The rules of inferences are modus ponens (stated above) and necessitation:

$$\text{If } \vdash \phi, \text{ then } \vdash \Box\phi$$

To get a complete axiomatization for the modal quantification theory, we combine the vocabulary, formation rules, axioms, and rules of inference for the propositional modal logic with those of the extensional quantification theory, but we need to add one further axiom that concerns the interaction of the modal operator with the quantifier, which I have labeled QCBF, the Qualified Converse Barcan Formula:

$$\vdash \Box \forall x \phi \rightarrow \forall x \Box (Ex \rightarrow \phi)$$

The axioms of the extensional quantification theory and the propositional modal logic can all be seen to be valid (true in the actual world of any model) in the semantics given below, and the rules of inference can be shown to preserve validity. The QCBF can also be shown to be valid, and its independence can be demonstrated by a variant semantics in which all the other axioms and rules are sound, but in which QCBF has countermodels.[5]

4.2.3. Modal semantics

There is a natural generalization of extensional semantics to modal semantics for quantification theory as well as for propositional logic. Propositions will be modeled set-theoretically in the standard way. The theory of propositions described in Chapter 2 ensures that the set of propositions forms a complete atomic Boolean algebra, and any such algebra can be modeled by the set of subsets of some given set of points. A standard Kripke model structure takes this form, where the points in the space for which the propositions are subsets are possible worlds. We will not be assuming that *every* set of possible worlds is a proposition. (The algebra of actual propositions might be a subalgebra of the set of points, but it will a sub-algebra that is itself complete and atomic.) The relation between the Kripke models and the intended interpretation of the modal language will need further discussion. At this point, our main concern is to provide a model theory

[5] See Stalnaker 1994.

that is sound[6] and complete for the logic, and that can therefore show the consistency of a first-order theory of propositions that has in its domain the propositions that its sentences express.

Following Kripke, we define a *model structure* as a triple $\langle W, D, \alpha \rangle$ where W is a non-empty set—the set of points or possible worlds. D is a function from W into sets of possible individuals, a domain for each possible world. For $w \in W, D_w$ is the domain of possible world w. We define U, the set of all possible individuals, as the union of D_w for all $w \in W$. The valuation function assigns *intensions* to the descriptive expressions, where an intension is a function from possible worlds into extensions, and extensions are as defined in the extensional theory. Assignment functions take variables into members of U, and the rule for assignment functions is the same as in the extensional theory: $[\![x]\!]^s{}_w = s(x)$. The semantic rules for the extensional operators are straightforward generalizations of the extensional rules: just add a subscript "*w*" to the valuation and domain functions in all the appropriate places. Here are the generalized rules:

If ϕ has the form $Ft_1 \cdot\cdot t_n$, then $v[\![\phi]\!]^s{}_w = 1$ iff $\langle [\![t_1]\!]^s{}_w, \ldots [\![t_n]\!]^s{}_w \rangle \in [\![F]\!]^s{}_w$. $[\![\phi]\!]^s{}_w = 0$ otherwise.

If ϕ has the form $\sim\psi$, then $[\![\phi]\!]^s{}_w = 1 - [\![\psi]\!]^s{}_w$

If ϕ has the form $(\psi_1 \wedge \psi_2)$, then $[\![\phi]\!]^s{}_w = [\![\psi_1]\!]^s{}_w \times [\![\psi_2]\!]^s{}_w$

If ϕ has the form $\forall F$, then $[\![\phi]\!]^s{}_w = 1$ if $[\![F]\!]^s{}_w = D_w$, and 0 otherwise.

If F has the form $\lambda x\psi$, then $[\![F]\!]^s{}_w = \{d \in D_w : [\![\psi]\!]^{s[d/x]}{}_w = 1\}$.

If ϕ has the form $\Box\psi$, then $[\![\phi]\!]^s{}_w = 1$ if $[\![\psi]\!]^s{}_{w^*} = 1$ for all $w^* \in W$,

$[\![\phi]\!]^s{}_w = 0$ if $[\![\psi]\!]^s{}_{w^*} \neq 1$ for some $w^* \in W$.

[6] I leave it to the reader to verify that the axioms are all valid, and that the rules preserve validity.

A *model* for the first-order modal language is a pair of a model structure and a valuation function: $\langle\langle W, D, \alpha\rangle, [\![\,]\!]\rangle$. A set of sentences is *satisfiable* iff there is a model in which all the sentences in the set are true in the actual world of the model, and a sentence is *valid* iff it is true in the actual world of all models.

Sentences, like all descriptive expressions, have semantic values that are functions from possible worlds to extensions, so since the extension of a sentence is a truth-value, their intensions are functions from possible worlds to 1 or 0, or equivalently, the set of possible worlds for which the sentence has the extension 1. The semantic rule for the necessity operator will be, in this context, equivalent to the rule for the necessity operator proposed in Chapter 2: If ψ expresses the necessary proposition, then $\Box\psi$ expresses the necessary proposition; if ψ expresses any other proposition, then $\Box\psi$ expresses the impossible proposition.

I will conclude this section with a remark about the treatment of singular terms in this version of first-order modal logic. Individual constants are unrestricted in the following sense: the semantics allows for terms of this kind that denote different individuals in different possible worlds. As a result, the abstraction/concretion principle for singular terms, $(\lambda y\phi t \leftrightarrow \phi^{t/y})$, is not a theorem, and is not valid, since it may move a term in or out of the scope of a sentential operator, and with some terms the scope difference may be semantically significant. But some individual constants may have constant functions as their values; that is, in Kripke's jargon, they may be rigid designators. Terms of this kind will satisfy a *rigidity condition* that is defined in the construction for the completeness theorem. A singular term t meets the rigidity condition in a possible world w of a model iff a sentence of the following form is true in that world:

$$(\mathrm{E}t \wedge \Box(\mathrm{E}t \rightarrow \lambda x\Box((\mathrm{E}x \vee \mathrm{E}t) \rightarrow x = t)t))$$

The singular terms that meet this condition do validate abstraction; that is, the following schema is valid in the semantics and is a theorem of the modal logic:

$$\text{T12:} \vdash \lambda x\Box((\text{E}x \vee \text{E}t) \rightarrow x = t)t \rightarrow (\lambda y\phi t \leftrightarrow \phi^{t/y}))$$

This theorem plays a crucial role in the completeness proof.

In addition, we can prove the universal generalization of the rigidity condition:

$$\text{T13:} \vdash \forall y(\text{E}y \wedge \Box(\text{E}y \rightarrow \lambda x\Box((\text{E}x \vee \text{E}y) \rightarrow x = y)y)$$

This theorem also plays a crucial role in the completeness proof.

The treatment of singular terms in the *extensional* quantification theory we began with follows the policy discussed in Chapter 1 of adopting a negative free logic. This is different from the policy recommended by Quine (eliminating singular terms altogether, but simulating them with predicates), but we argued in Chapter 1 that it is essentially equivalent to it, and it is a policy that is like Quine's in that it yields a logic that is ontologically neutral, maintaining the division of labor between the commitments of the logic and the commitments of the choice of a lexicon for a regimented language. This treatment of singular terms in the modal semantics results from the natural generalization of extensions to intensions, according to which the semantic value of each item of the primitive descriptive vocabulary is a function from possible worlds to an extension of the appropriate kind.

This is the way the particular semantics I am developing works, but there are alternative decisions one might have made in choosing how to do the semantics, decisions that do not involve any different substantive commitments. I will mention two alternatives: First, one might have followed Quine's policy by restricting our language to a *pure* modal quantification theory with no singular terms other than variables.[7] Second, one might instead allow only a more restricted class of singular terms, for example allowing only terms that are each stipulated to apply to a particular individual. This decision would result in what Bruno Jacinto called a "strongly Millian modal logic," allowing singular terms, but building the rigidity condition on names into the logic, and then presupposing that singular terms are *admissible* to the regimented language only if they succeed in referring.[8] I want to make three comments about the choices between the strategy we are following and these two alternatives: First, the choice calls for a decision, but not a judgment about what commitments it is appropriate to make or to presuppose. The basic model structure is the same on any of these choices. Furthermore, while the Millian alternative weakens the expressive power of the language, in a sense, by banning a kind of singular term (non-rigid designators) that can be perfectly well-defined with the resources of the model theory, this restriction, in another sense, does not weaken the expressive power of the language since one can always simulate non-rigid singular terms with predicates, following the maneuver that Quine used. One might, however, think that the simulation maneuver distorts the logical form of statements that are more naturally expressed with non-rigid designators such as definite descriptions. Furthermore, the restriction to rigid terms does not add anything, since rigid designators are allowed as a special case in our less restrictive language, and they can be identified and distinguished within that language.

Second comment: the Millian alternative requires that one distinguish, as Jacinto does in his discussion, between a sense of validity (which he calls "real-world validity") that is not preserved by the rule of necessitation and a sense that he calls "general validity" that requires not just truth in the actual world of all admissible

[7] The quantified modal logic defined in Kripke 1963 followed Quine's policy.
[8] See Jacinto 2017.

models, but truth in all possible worlds of all admissible models. Real-world validity is truth in the actual world of all models on the assumption that all of the primitive terms of the descriptive vocabulary of the language are admissible, where admissibility is admissibility in the actual world. So, for example, in the Millian semantics, a sentence of the form $\exists xx = t$ (where t is any singular term) will be real-world valid, since an individual constant t will be admissible only if it refers to something, but the constraints on the admissibility of terms in the descriptive lexicon of a regimented language are constraints on the actual world—the world in which the language is being defined. So a name will be admissible in a Millian language only if it succeeds in picking out an individual. But this constraint is compatible with the individual it picks out being a contingent existent. So the necessitation of the existential statement might be false in the actual world, and so not real-world valid. The divergence between the two notions of validity results from the fact that the Millian logic builds some of the substantive presuppositions of a regimented language/theory into the logic, and so disrupts the sharp division of labor that Quine was aiming at between the commitments imposed by logic and the commitments imposed by the choice of a lexicon. But the notion of real-world validity is perfectly well defined, and the notion of *general* validity does respect the Quinean aim of ontological neutrality.

The third comment is the observation that when we generalize beyond first-order logic, as we will do in the next chapter, a form of the problem that the Millian logic raises will arise for predicates, and not just for names, and this problem is less easily avoided. The judgment that a *predicate* is admissible may also have contingent presuppositions, and if our language can quantify over the semantic values of predicates (properties and relations), then some quantified statements of this kind may be real-world valid even if their necessitations are not. But discussion of this issue will be postponed to Chapter 5.

4.3. A model for a theory of propositions

In this section, we shift gears to consider the application of the model theory to the special case where the language to be interpreted in the model theory is our theory of propositions. That is, the predicates of the language will be the predicates of our first-order theory of propositions, and the domain of the model will be a set of propositions. The construction of a model for this purpose will build on the general completeness result (proved in Appendix II), and on the construction defined in the proof of that result. That proof provided the construction of a model to satisfy any given arbitrary consistent set of sentences, and the particular model structure we construct will have certain general features not shared by all models (e.g., there will be a denumerable number of possible worlds), and so there will be corollaries of the completeness theorem that for any arbitrary set of sentences of the first-order language, there exists a model that meets certain additional conditions. In this section of this chapter, we will build on the construction used for completeness to state a corollary of this kind. We will then begin with an arbitrary model meeting the conditions specified in the corollary and define in terms of it a second model for a first-order theory of propositions, where the predicates are those of first-order theory of propositions stated in Chapter 2, and the propositions that are the subject matter of that theory (the members of its domain) are the propositions expressible in the language of the first model. In the second model we define, all the propositions *expressible* in the theory of propositions will be among the propositions in the domain of that theory. That is, our second theory will be a theory about the propositions that are expressed in that same theory.

In the model that is defined for the completeness proof, possible worlds are identified with non-negative integers, and for each possible world k of the model, a set of sentences Γ_k is constructed to represent the world by containing all of the sentences of an augmented language that are true in that possible world. Before stating the corollary to the completeness theorem that I will be building on, I will streamline the constructed model in two ways that do not affect the result. First, the construction allows for the possibility that there are distinct possible worlds that determine the same set of sentences.[9] That is there might be a *j* and *k* such that $j \neq k$, but $\Gamma_j = \Gamma_k$. It will be convenient to prune the model by eliminating all but one of the integers that are equivalent in this way, ensuring that our model is a *distinguished* model: one for which there is a one-one correspondence between worlds and sets of truths of the language. Second, the saturated sets that represent the truths is a world in which there are sets of sentences of the language L^+, which is the original language augmented by a denumerable set **R** of extra names. In the definition of the model, a subset $\mathbf{R}^*$ of **R** was specified as the set of all possible individuals (the union of the domains of the worlds). It will be convenient to drop from our language all of the extra names in **R** except for those in $\mathbf{R}^*$. The model streamlined in these two ways will still satisfy all the conditions needed for the completeness argument.

Using these sets of sentences, we can define an *equivalence system* for the possible worlds: this is a system of equivalence relations, one such relation for each of the worlds, or equivalently a three place relation on worlds, $j \approx_i k$, where each '$\approx_i$' is an equivalence relation. Furthermore, we can show that this system is a *coherent* equivalence system, in a sense to be defined.

Before getting to the formal definition of coherence and the argument that the model we have constructed determines a coherent equivalence system, I will make some remarks about the role that this notion will play and its intuitive motivation. The propositions of a model structure are represented by sets of possible worlds, but we don't need to assume that every set of possible worlds is a proposition. It may be that some distinct non-actual possible worlds are

[9] Thanks to Peter Fritz (personal communication) for pointing this possibility out to me.

equivalent in the sense that they have the same representational significance from the perspective of the actual world. Equivalent worlds are worlds that are indistinguishable with the resources for distinguishing possibilities that are available in the actual world. But each possible world is a potential actual world, and we need to consider, for a counterfactual world *w*, the resources that *would* be available for distinguishing between possibilities if *w* were actual. So it may be a contingent fact that two possible worlds are equivalent in this sense, and so we need a world-dependent equivalence relation, and that is what an equivalence system provides.

Recall that the Quinean conception of proposition that we started with in Chapter 2—the idea of objective information—was defined relative to a matrix of alternatives that might have substantive presuppositions. On the naturalistic Quinean picture that rejects the possibility of an a priori transcendent standpoint, we must assume that the means available to distinguish between alternative possibilities depend on facts about the actual world in which those distinctions are being made. So our characterization of alternative possibilities will recognize that if the facts had been different, our resources for distinguishing between possibilities might have been different, and as a result different propositions would have existed: if the facts had been different, some distinctions unavailable from the perspective of the actual world might have been available, and some distinctions that *are* available from the perspective of the actual world might not have been available. Since our complete characterization of logical space must be given from the perspective of the actual world, what we can say about the distinctions that we cannot make, but that we would be able to make if things were different, will be limited. As illustrated with examples in Chapter 3, we can talk in a general way about possibilities in which there would exist things other than those that do exist (Saul Kripke might have had seven sons), but not (or not necessarily) about particular things that might have existed (a particular seventh son), but do not. Propositions describing counterfactual situations might be maximal in the sense that they make all the distinctions that can be made from our perspective, but that still fail to be fully specific. We can, however, talk at one remove, in a general way, about the distinctions that could have been made if things had been different in certain respects. The aim of the definition of a coherent equivalence system on the possible worlds of a Kripke model is to characterize at a high level of abstraction the relationships between the distinctions that can be made from the perspectives of different possible situations. To say that two possible worlds, j and k, of a Kripke structure are equivalent from the perspective of a third possible world i is to say that the resources available in world i are not able to distinguish j from k. This will imply that the distinctions between possibilities that can be made from the perspective of j must be structurally similar to the distinctions that can be made from the perspective of k. Otherwise we could distinguish j from k in terms of their structural differences. The notion of coherence is a way of making the relevant structural constraints precise. We will have more to say in Chapter 6 about the motivation and application of the equivalence systems, and the relationship between Kripke models and the reality that they purport to be modeling. For now, I will just spell out the formal notion of a coherent equivalence system, and the argument that the model that is constructed for the completeness theorem determines a coherent equivalence system for the propositions defined for that model. The upshot will be the following corollary of the completeness theorem:

> *For every consistent set of sentences of the modal language, there exists a model that determines a coherent equivalence system of propositions.*

One further remark before getting to the details: In the construction for the completeness theorem, we augmented the language by adding vocabulary—singular terms that were stipulated to satisfy a condition that required them to be, in the model defined, rigid designators for individuals that exist in the various possible worlds. Some of these rigid names refer to things in the domain of the actual world; these are names whose referents might be fixed by the individuals themselves, and so would be admissible in the language of the actual world, but names for individuals that would exist if things were different are not necessarily admissible in the language of the actual world, since there will exist no individuals to fix the reference of those names. This implies that sentences containing those names may not express propositions that exist in the actual world. Each possible world of the model is, in a sense, a potential actual world, and the names that were introduced to represent members of the domain of a counterfactual world would be admissible if that world were actual. While the augmented language L^+ of our streamlined model for the construction in the completeness argument contains all of the names in R^* that are in the domains of any of the possible worlds, we might define, a sublanguage L_k of L^+ for each possible world k as the language that excludes those members of R^* that do not refer in that world (those $r \in R^*$ such that $\exists x x = r \notin \Gamma_k$). Each of the saturated sets, Γ_k, remains saturated when we restrict its members to sentences in the language L_k, and with this restriction, we will be able to show that every sentence of L_k expresses a proposition that exists in world k.[10]

An equivalence system (a three-place relation taking a possible worlds i, to an equivalence relation, $\approx_i$) is *coherent* iff it meets two conditions, a fixed-point condition and a structure-preserving condition. The first condition is that each possible world is equivalent (relative to that world) only to itself: $i \approx_i j$ iff $j=i$. This condition

[10] Any individual constant that is in the original language L will remain in all of the languages L_k. Such names might satisfy the rigidity condition, or not. They might even satisfy, in the actual world, the modal part of the rigidity condition without satisfying the existential part. In this last case these names will express (in the model) individual essences that are not instantiated. The contingentist metaphysics will not allow for names or predicates whose reference is fixed by merely possible individuals, but it does not exclude the admissibility of some predicates or individual constants that identify uninstantiated individual essences.

is motivated by the assumption that the actual world contains the resources to distinguish the actual world from others. The second condition is that any two worlds j and k that are equivalent from the perspective of a world i must be isomorphic with respect to the structure of distinctions that can be made from their perspectives. This requirement is made precise in the following way: for any j and k such that $j \approx_i k$, there must be an admissible structure preserving permutation function f such that $f(j) = k$. A permutation function is admissible iff for all i, j and k, such that $j \approx_i k$, $f(j) \approx_i k$, and it is structure preserving (an automorphism) iff for all i, j and k, $j \approx_i k$ iff $f(j) \approx_{f(i)} f(k)$.[11]

The definition of the particular equivalence system for our streamlined model will be indirect. We will first define what Peter Fritz called a *permutation system* for the possible worlds of that model. This is a set of sets of permutations[12] of the set of possible worlds, each indexed to one of the possible worlds. In terms of any permutation system we can define a corresponding equivalence system. Some permutation systems satisfy a condition that corresponds to the coherence condition for equivalence systems, and it can be shown that permutation systems that satisfy this condition determine coherent equivalence systems.

The sentences that characterize the possible worlds in the model constructed for the completeness theorem were in a language $\mathbf{L}^+$ that contained a set $\mathbf{R}^*$, which is both a set of rigid designators of members of the domains of the different possible worlds and are

[11] I defined coherent equivalence systems for propositions in Appendix A in Stalnaker 2012, though I did not use this terminology there. My treatment in that appendix was overly compressed, and contained a number of errors that were pointed out and corrected in Fritz 2016. I am here adopting Fritz's terminology and incorporating his corrections to my original definition. In addition to correcting the errors, Fritz's excellent paper also gives a more systematic account of the mathematical structure of coherent equivalence systems, illustrated with simple models that helped to make it clear. And even though he is critical of the contingentist metaphysics for a number of reasons, he gives a clear and sympathetic account of its intuitive motivation.

[12] One-one functions from a set onto itself.

the referents of those names.[13] Our construction of a permutation system starts with the set of all permutations on these names. A permutation on names will determine a permutation of sentences of the language in the following way: If f is the permutation of names, then for any sentence ϕ, $f(\phi)$ is the result of substituting $f(a)$ for a for all a in ϕ. It will also determine a permutation of *sets* of sentences: for any set of sentences Γ, $f(\Gamma) = \{f(\phi) : \phi \in \Gamma\}$. In the streamlined version of our construction for the completeness theorem, the possible worlds of the model correspond 1-1 with saturated sets of sentences: world k corresponds to the set of sentences Γ_k, for each world k. For some of the permutations f, it will be true that for each of the saturated sets Γ_k, $f(\Gamma_k)$ will also be a saturated set Γ_j, for some j. Call permutations meeting this conditions "closed." (We know that there will be at least one such permutation, since the identity function is a closed permutation function.) Now we can define a set of permutations F_k, for each world k, as follows: $f \in F_k$ iff (1) f is closed, and (2) $f(c) = c$ for all $c \in \mathbf{R}^*$ such that $Ec \in \Gamma_k$. Since all of these permutations are closed, they determine a system of permutations of the possible worlds of our model.

We say that a permutation system is *coherent* iff it satisfies these three conditions: (1) if $f \in F_x$, then $f(x) = x$; (2) for each x, F_x is closed under inverse and composition; (3) if $f \in F_x$ and $g \in F_y$, then $fgf^* \in F_{f(y)}$ (where f^* is the inverse of f, and fgf^* is the composition of the three permutation functions. For any permutation system, we can define the equivalence relations for the corresponding equivalence system as follows: $y \approx_x z$ iff for some $f \in F_x, f(y) = z$. It can then be shown, first, that the permutation system we have defined satisfies its coherence condition, and

[13] The names in this set that are not in the domain of the actual world represent things that don't exist, and of course it is a fiction that the referents of these names (which are the names themselves) do not exist. There are no nonexistent things, but to model the possibility that there are things that do not in fact exist, we need artifacts of the model that behave like names but do not name anything in the reality being modeled. See Chapter 6 for some discussion of this issue.

second that an equivalence system defined in terms of a permutation system in this way satisfies its coherence condition.[14]

We are now in a position to use the streamlined model we have constructed, which was a model for an arbitrary consistent set of sentences of the modal language, to construct a model for the theory of propositions. Each proposition is modeled by a set of possible worlds, but not every set of possible worlds will represent a proposition that exists in a given possible world, since worlds that are equivalent with respect to world w_i will be indistinguishable with respect to the resources available in world w_i. Equivalence classes defined by the relation $\approx_i$ will be propositions that are maximal with respect to world *i*, and the propositions that exist in world *i* will be all the sets of possible worlds that are unions of these maximal propositions. And it is straightforward to show that for each world *k*, every sentence of the language L^k expresses a proposition that exists in world *k*. This must be true because if ϕ is a sentence of L^k, it can contain only rigid names that refer in *k*, and so for any permutation *f* in F_k and any world *j*. $\phi \in \Gamma_j$ iff $\phi \in \Gamma_{f(j)}$.

This completes the argument for the corollary of the completeness theorem: that every consistent set of sentences of the first-order modal language has a Kripke model that determines a coherent equivalence system, a system that determines a set of propositions for each possible world. We are now in a position to use the structure of propositions determined by any model meeting this condition to define a first-order model for our theory of propositions, where the propositions expressed in each of the possible worlds of the first model are the proposition in the domain of the theory.

In a model for the simple extensional first-order theory of propositions that we defined in Chapter 2, the domain consisted

[14] The coherence condition for permutation systems is spelled out in Stalnaker 2012 (Appendix A) and in Fritz 2016. An argument that the permutation system defined here meets the coherence condition is a minor modification of an argument given (for a slightly different conclusion) in Stalnaker 2012 (Appendix C).

of both propositions and sets of them, and one of the two primitive predicate of the theory ('C' for consistency) applied both to propositions and to sets of them. So in our Kripke model for the modal generalization of this theory, the domains should contain both propositions (certain sets of possible worlds) and sets of them (sets of all the sets of possible worlds that are the propositions that exist in that world. But for this exposition I will oversimplify by defining a model with just propositions in the domains. According to the theory, every set of propositions corresponds to a proposition, and equivalent sets of propositions (sets that correspond to the same proposition) will be indistinguishable by the predicate of consistency, so nothing important is lost in this simplification. The defined predicates (entailment, contradictory, etc.) can all be defined in the meta-language in terms of consistency and truth.

In the model structure $\langle W, w_0, D \rangle$ for our theory of propositions, the set of worlds, W and the actual world, w_0, are the same as for the original model, but the domain function is different: D_w is the set of propositions that exist in world w. The vocabulary of the theory has two primitive predicates, 'T' (truth) and 'C' (consistency). Here are the specifications of the intensions (functions from possible worlds to extensions) for these two predicates:

$$[\![T]\!]_i = \{d \in D_i : w_i \in d\}$$

$$[\![C]\!]_i = \{d \in D_i : d \text{ is nonempty}\}$$

The postulates of the theory can all be seen to be necessarily true, given these specifications.

So what does this construction establish? We have shown first, building on the completeness theorem, that an arbitrary consistent set of sentences of the modal language has a first-order Kripke model that determines a coherent equivalence system of propositions, and second, that there is a Kripke model for the first-order theory of propositions, where the propositions that are the subject matter of the theory are those that are determined by this first model structure.

The propositions *expressed* in the theory of propositions are among those in the domain of the theory. So the construction shows that we can define a model for the theory that describes and quantifies over the propositions that it expresses. Among the propositions both expressed and described in that theory are propositions that ascribe the property of truth to propositions.

We have two parallel models, two regimented modal languages with the same set of possible worlds, but with different ontologies. Can we put the two parallel models together into a single model with entities of the different kinds in a single first-order domain for each possible world? I see no reason why not, though the merging of different frameworks that involve different ontological commitments will be a further development—a further project of regimentation requiring further decisions and commitments. The predicates of both of the parallel theories are defined only for their restricted domains, so if we merge the two frameworks, we need to say more about the relationships between the entities in the two domains, and about the application of the predicates of each of the theories to the entities in the domain of the other. We might merge the two domains and add a predicate applying to all and only the propositions. It might be that the type-one individuals (the individuals in the domains of the first theory) contain no propositions, and that the primitive descriptive predicates of the two sub-theories are stipulated to apply only to things in the separate subdomains. (So, for example, if there are dogs and cats among the type-one individuals, these individuals will be neither true nor false, and neither consistent nor inconsistent, and none of the propositions will be Doberman pinschers or Manx cats.) In the merged formulation, the typed quantifiers will become defined restricted quantifiers, ranging over propositions and non-propositions, respectively. Or alternatively, it might be that the original theory already contained a predicate, applying to all and only propositions, and a version of the theory of propositions as a sub-theory. Either way, the first-order formulation with a single domain for each world might be augmented with new predicates expressing relationships between things in the two subdomains. But I will set aside for now questions about the merging of the two first-order theories. In the next chapter when we explore the hypothesis that there are properties and relations as well as propositions, we will consider a different framework (a higher-order language) for describing the relationships between entities of different types.

We have been talking in this chapter about Kripke models, and there is much more that needs to be said about the reality that a model with this structure might be modeling. In the semantic meta-language, we talk freely about merely possible individuals, and merely possible propositions, even though we say that there are not such things in the reality being modeled. But I think the constructions show that there is no formal problem—no paradox—in the idea of a theory of propositions that is about the propositions expressed in that theory.

Appendix I
Some Theorems of the System

Thirteen theorem schemata are listed, with sketches of their proofs from the axioms. This includes all of the theorems involving quantifiers that are appealed to in the completeness proof. Note that eleven of the thirteen theorems use only axioms of the extensional predicate logic.

These are proof sketches rather than fully explicit proofs, compressing some steps. For example, the quantifier distribution axiom $(\forall x(\phi \rightarrow \psi) \rightarrow (\forall x\phi \rightarrow \forall x\psi)$ easily generalizes to distribution over successive conditionals, over the biconditional, and over conjunction, so we will take these for granted in appealing to Q distribution.

Proofs are sequences of *closed* sentences, so all occurrences of 's', 't', and so on in the proof schemata will take individual constants (and not variables) as their values.

T1 $\vdash \exists x x = t \rightarrow t = t$

1. $s = t \rightarrow (s = t \rightarrow t = t)$	Identity axiom
2. $t \neq t \rightarrow s \neq t$	1, TF
3. $t \neq t \rightarrow \forall x(x \neq t)$	2, UG
4. $\exists x x = t \rightarrow t = t$	3, TF

T2 $\vdash \forall y \exists x x = y$

1. $(\lambda z(Fz \vee \sim Fz)t) \leftrightarrow (Ft \vee \sim Ft)) \rightarrow \lambda z(Fz \vee \sim Fz)t$	TF
2. $\lambda z(Fz \vee \sim Fz)t \rightarrow \exists x x = t$	Existence axiom
3. $(\lambda z(Fz \vee \sim Fz)t \leftrightarrow (Ft \vee \sim Ft)) \rightarrow \exists x x = t$	1, 2, TF
4. $\forall y((\lambda z(Fz \vee \sim Fz)y \leftrightarrow (Fy \sim Fy)) \rightarrow \exists x x = y)$	3, UG
5. $\forall y(\lambda z(Fz \vee \sim Fz)y \leftrightarrow (Fy \vee \sim Fy)) \rightarrow \forall y \exists x x = y$	4, Q distribution
6. $\forall y(\lambda z(Fz \vee \sim Fz)y \leftrightarrow (Fy \vee \sim Fy))$	Abstraction axiom
7. $\forall y \exists x x = y$	5,6, TF

T3 $\vdash \forall x x{=}x$

1. $\exists y y = t \rightarrow t = t$	T1
2. $\forall x(\exists y y = x \rightarrow x = x)$	1, UG
3. $\forall x \exists y y = x \rightarrow \forall x x = x$	Q distribution
4. $\forall x \exists y y = x$	T2
5. $\forall x x = x$	3,4, MP

T4 $\vdash s = t \rightarrow t = s$

1. $\exists x x = t \rightarrow t = t$	T1
2. $s = t \rightarrow (t = t \rightarrow t = s)$	Identity axiom
3. $s = t \rightarrow \exists x x = t$	Existence axiom
4. $s = t \rightarrow t = s$	1,2,3, TF

T5 $\vdash \forall F \rightarrow (\exists x x = t \rightarrow Ft)$

1. $s = t \rightarrow (Fs \rightarrow Ft)$	Identity axiom
2. $\sim Ft \rightarrow (Fs \rightarrow s \neq t)$	1, TF
3. $\sim Ft \rightarrow \forall x(Fx \rightarrow x \neq t)$	2, UG
4. $\sim Ft \rightarrow (\forall x Fx \rightarrow \forall x x \neq t)$	3, Q distribution
5. $\forall x Fx \rightarrow \forall F$	second Q axiom
6. $\forall F \rightarrow (\exists x x = t \rightarrow Ft)$	4,5, TF

T6 $\vdash \forall y(\forall x \phi \rightarrow \phi^{y/x})$

1. $\forall x \phi \rightarrow (\exists x x = t \rightarrow \lambda x \phi t)$	T5
2. $(\lambda x \phi t \leftrightarrow \phi^{t/x}) \rightarrow (\exists x x = t \rightarrow (\forall x \phi \rightarrow \phi^{t/x}))$	1, TF
3. $\forall y((\lambda x \phi y \leftrightarrow \phi^{y/x}) \rightarrow (\exists x x = y \rightarrow (\forall x \phi \rightarrow \phi^{y/x})))$	2, UG
4. $\forall y(\lambda x \phi y \leftrightarrow \phi^{y/x}) \rightarrow (\forall y \exists x x = y \rightarrow \forall y$ $(\forall x \phi \rightarrow \phi^{y/x}))$	3, Q distribution
5. $\forall y(\lambda x \phi y \leftrightarrow \phi^{y/x})$	Abstraction axiom
6. $\forall y \exists x x = y$	T2
7. $\forall y(\forall x \phi \rightarrow \phi^{y/x}))$	4,5,6, TF

T7 $\vdash \forall x \phi \rightarrow \forall y \phi^{y/x}$

1. $\forall y(\forall x \phi \rightarrow \phi^{y/x})$	T6
2. $\forall y \forall x \phi \rightarrow \forall y \phi^{y/x}$	1, Q distribution

3. $\forall x\phi \to \forall x\phi$	TF
4. $\forall x\phi \to \forall y\forall x\phi$	3, UG
5. $\forall x\phi \to \forall y\phi^{y/x}$	2,4,TF

T8 $\vdash \exists xx=t \leftrightarrow \lambda y\exists xx=yt$

The right-to-left direction is immediate by the existence axiom, so we need only to prove the left-to-right direction.

1. $s=t \to (\lambda y\exists xx=ys \to (\lambda y\exists xx=yt)$	Identity axiom
2. $\sim\lambda y\exists xx=yt \to (\lambda y\exists xx=ys \to s\neq t)$	1, TF
3. $\sim\lambda y\exists xx=yt \to \forall z(\lambda y\exists xx=yz \to z\neq t)$	2, UG
4. $\sim\lambda y\exists xx=yt \to (\forall z\lambda y\exists xx=yz \to \forall zz\neq t)$	3, Q distribution
5. $\exists zz=t \to (\forall z\lambda y\exists xx=yz \to \lambda y\exists xx=yt)$	4, TF
6. $\forall z(\lambda y\exists xx=yz \leftrightarrow \exists xx=z)$	Abstraction axiom
7. $\forall z\lambda y\exists xx=yz \leftrightarrow \forall z\exists xx=z)$	6, Q distribution
8. $\forall z\exists xx=z$	T2
9. $\exists zz=t \to \lambda y\exists xx=yt$	5,7,8, TF
10. $\exists zz=t \to \exists zz=t$	T7
11. $\exists zz=t \to \lambda y\exists zz=yt$	9,10 TF

T9 $\vdash \forall x(\phi \to \psi) \to (\lambda x\phi s \to \lambda x\psi s)$

1. $(\lambda x\phi t \leftrightarrow \phi^{t/x}) \to ((\lambda x\psi t \leftrightarrow \psi^{t/x}) \to ((\phi^{t/x} \to \psi^{t/x})$ $\to (\lambda x\phi t \to \lambda x\psi t)))$	TF
2. $s=t \to ((\lambda x\phi t \to \lambda x\phi s) \wedge (\lambda x\psi t \to \lambda x\psi s)$	Identity axiom, TF
3. $t=s \to ((\lambda x\phi s \to \lambda x\phi t) \wedge (\lambda x\psi ts \to \lambda x\psi t)$	Identity axiom, TF
4. $s=t \to t=s$	T4
5. $s=t \to ((\lambda x\phi t \leftrightarrow \lambda x\phi s) \wedge (\lambda x\psi t \leftrightarrow \lambda x\psi s)$	2,3,4,TF
6. $\sim(\lambda x\phi s \to \lambda x\psi s) \to ((\lambda x\phi t \leftrightarrow \phi^{t/x}) \to ((\lambda x\psi t \leftrightarrow \psi^{t/x})$ $\to ((\phi^{t/x} \to \psi^{t/x}) \to s\neq t)))$	1,5 TF
7. $\sim(\lambda x\phi s \to \lambda x\psi s) \to \forall x((\lambda x\phi x \leftrightarrow \phi) \to ((\lambda x\psi x \leftrightarrow \psi)$ $\to ((\phi \to \psi) \to s\neq x)))$	6, UG
8. $\sim(\lambda x\phi s \to \lambda x\psi s) \to \forall x((\lambda x\phi x \leftrightarrow \phi)$ $\to (\forall x(\lambda x\psi x \leftrightarrow \psi) \to (\forall x(\phi \to \psi) \to \forall xx\neq s)))$	7, Q distribution
9. $\forall x(\lambda x\phi x \leftrightarrow \phi$	Abstraction axiom
10. $(\forall x(\lambda x\psi x \leftrightarrow \psi)$	Abstraction axiom
11. $\exists xx=s \to ((\forall x(\phi \to \psi) \to (\lambda x\phi s \to \lambda x\psi s))$	8,9,10, TF
12. $\lambda x\phi s \to \exists xx=s$	Existence axiom
13. $(\forall x(\phi \to \psi) \to (\lambda x\phi s \to \lambda x\psi s)$	11,12 TF

T10 $\vdash \Box(\mathrm{E}s \vee \mathrm{E}t) \to s=t) \to (\phi \leftrightarrow \phi^{t/s})$

For each of the theorem schemata before this we have sketched a particular proof schema—a sequence of sentence schemata where each sequence of instances of those sentence schemata is a proof. In contrast, this argument will proceed by showing, by induction, how to construct a proof for any instance of the schema, but the particular proofs will depend on the complexity of the sentence ϕ. We first show that T10 holds for the case where ϕ is an atomic sentence, and then show that if the schema is a theorem for all sentences ϕ of length k or less, then it is a theorem for all ϕ of length k+1.

The base case: Suppose ϕ is an atomic sentence of the form $\mathrm{R}t_1, \ldots t_n$, where R is a primitive predicate. If s does not occur in ϕ, then the claim is a tautology, so assume that s occurs in ϕ.

1. $\Box(\mathrm{E}s \vee \mathrm{E}t) \rightarrow s = t) \rightarrow ((\mathrm{E}s \vee \mathrm{E}t) \rightarrow s = t)$ S5 axiom $(\psi \rightarrow \psi)$
2. $s = t \rightarrow (\phi \rightarrow \phi^{t/s})$ Identity axiom
3. $s = t \rightarrow t = s$ T4
4. $t = s \rightarrow (\phi^{t/s} \rightarrow \phi)$ Identity axiom
5. $s = t \rightarrow (\phi \rightarrow \phi^{t/s})$ 2,3,4, TF
6. $\Box(\mathrm{E}s \vee \mathrm{E}t) \rightarrow s = t) \rightarrow ((\mathrm{E}s \vee \mathrm{E}t) \rightarrow (\phi \rightarrow \phi^{t/s}))$ 1,5 TF
7. $\phi \rightarrow \mathrm{E}s$ Existence axiom
8. $\phi^{t/s} \rightarrow \mathrm{E}t$ Existence axiom
9. $\sim(\mathrm{E}s \vee \mathrm{E}t) \rightarrow (\phi \rightarrow \phi^{t/s})$ 7,8, TF
10. $\Box(\mathrm{E}s \vee \mathrm{E}t) \rightarrow s = t) \rightarrow (\phi \leftrightarrow \phi^{t/s}))$ 6,9, TF

Now for the induction steps, we consider each kind of complex sentence, $\phi = \sim\psi$, $\phi = (\psi_1 \wedge \psi_2)$, $\phi = \Box\psi$, $\phi = \forall\lambda x\psi$, $\phi = \lambda x\psi t$. In each case, if ϕ has k+1 symbols, each of the ψ sentences has k or fewer symbols, so by hypothesis of induction, the claim holds for them. This step is obvious for the truth-functional operators.

For the case where $\phi = \Box\psi$:

1. $\Box((\mathrm{E}s \vee \mathrm{E}t) \rightarrow s = t) \rightarrow (\psi \leftrightarrow \psi^{t/s})$ hypothesis of induction
2. $\Box\Box((\mathrm{E}s \vee \mathrm{E}t) \rightarrow s = t) \rightarrow (\Box\psi \leftrightarrow \Box\psi^{t/s})$ 1, necessitation and $\Box$ distribution
3. $\Box((\mathrm{E}s \vee \mathrm{E}t) \rightarrow s = t) \rightarrow \Box\Box(\mathrm{E}s \vee \mathrm{E}t) \rightarrow s = t$ S5 theorem $(\Box\phi \rightarrow \Box\Box\phi)$
4. $\Box((\mathrm{E}s \vee \mathrm{E}t) \rightarrow s = t) \rightarrow (\Box\psi \leftrightarrow \Box\psi^{t/s})$ 2,3, TF

For the case where $\phi = \forall x\psi$: Let r be an individual constant other than s or t.

1. $\Box((\mathrm{E}s \vee \mathrm{E}t) \rightarrow s = t) \rightarrow (\psi^{r/x} \leftrightarrow \psi^{r/x\ t/s})$ hypothesis of induction
2. $\Box((\mathrm{E}s \vee \mathrm{E}t) \rightarrow s = t) \rightarrow \forall x(\psi \leftrightarrow \psi^{t/s})$ 1, UG
3. $\Box((\mathrm{E}s \vee \mathrm{E}t) \rightarrow s = t) \rightarrow (\forall x\psi \leftrightarrow \forall x\psi^{t/s})$ 2, Q distribution

Finally, for the case where $\psi = \lambda x\psi(c)$: Let r be an individual constant different from c, t, or s:

1. $\Box((\mathrm{E}s \vee \mathrm{E}t) \rightarrow s = t) \rightarrow (\psi^{r/x} \leftrightarrow \psi^{r/x\ t/s})$ hypothesis of induction
2. $\Box((\mathrm{E}s \vee \mathrm{E}t) \rightarrow s = t) \rightarrow \forall x(\psi \leftrightarrow \psi^{t/s})$ 1, UG
3. $\Box((\mathrm{E}s \vee \mathrm{E}t) \rightarrow s = t) \rightarrow \lambda x\psi c \leftrightarrow \lambda x\psi^{t/s}c)$ 2, T9

If c is different from s, then $\lambda x\psi^{t/s}c$ is the same as $\lambda x\psi c^{t/s}$, so we have our result. But in the special case where c is the same as s, step 3 will take the form 3', and we can complete the proof as follows:

3'. $\Box((\mathrm{E}s \vee \mathrm{E}t) \rightarrow s = t) \rightarrow (\lambda x\psi s \leftrightarrow \lambda x\psi^{t/s}s)$
4'. $\Box((\mathrm{E}s \vee \mathrm{E}t) \rightarrow s = t) \rightarrow (\lambda x\psi^{t/s}s \leftrightarrow \lambda x\psi^{t/s}t)$ — as shown in the proof of the base case
5'. $\Box((\mathrm{E}s \vee \mathrm{E}t) \rightarrow s = t) \rightarrow (\lambda x\psi s \leftrightarrow \lambda x\psi^{t/s})$ — 3',4', TF

This completes the proof of T10.

T11 $\vdash \lambda x \sim \phi t \rightarrow \sim \lambda x\phi t$

1. $t = s \rightarrow (\lambda x\phi t \leftrightarrow \lambda x\phi s)$ — Identity axiom
2. $t = s \rightarrow (\lambda x \sim \phi t \leftrightarrow \lambda x \sim \phi s)$ — Identity axiom
3. $\sim(\lambda x \sim \phi t \rightarrow \lambda x\phi s) \rightarrow ((\lambda x\phi t \leftrightarrow \phi^{t/x})$ $\rightarrow ((\lambda x \sim \phi t \leftrightarrow \sim \phi^{t/x}) \rightarrow t \neq s))$ — 1,2,TF
4. $\sim(\lambda x \sim \phi s \rightarrow \sim \lambda x\phi s) \rightarrow \forall x((\lambda x\phi x \leftrightarrow \phi)$ $\rightarrow ((\lambda x \sim \phi x \leftrightarrow \sim \phi) \rightarrow x \neq s))$ — 3,UG
5. $\sim(\lambda x \sim \phi s \rightarrow \sim \lambda x\phi s) \rightarrow \forall x((\lambda x\phi x \leftrightarrow \phi)$ $\rightarrow (\forall x(\lambda x \sim \phi x \leftrightarrow \sim \phi) \rightarrow \forall x x \neq s))$ — 4,Q distribution
6. $\forall x(\lambda x\phi x \leftrightarrow \phi)$ — Abstraction axiom
7. $\forall x(\lambda x \sim \phi x \leftrightarrow \sim \phi)$ — Abstraction axiom
8. $\exists x x = s \rightarrow (\lambda x \sim \phi x \rightarrow \sim \lambda x\phi s)$ — 5,6,7,TF
9. $(\lambda x \sim \phi s \rightarrow \exists x x = s)$ — Existence axiom
10. $(\lambda x \sim \phi s \rightarrow \sim \lambda x\phi s)$ — 8,9, TF

T12 $\vdash \lambda x\Box((\mathrm{E}x \vee \mathrm{E}t) \rightarrow x = t)t \rightarrow (\lambda x\phi t \leftrightarrow \phi^{t/x})$

1. $\Box((\mathrm{E}s \vee \mathrm{E}t) \rightarrow s = t) \rightarrow (\phi^{s/x} \leftrightarrow \phi^{t/x})$ — T10
2. $\phi^{t/x} \rightarrow \Box((\mathrm{E}s \vee \mathrm{E}t) \rightarrow s = t) \rightarrow \phi^{s/x})$ — 1, TF
3. $\phi^{t/x} \rightarrow \forall x\Box((\mathrm{E}s \vee \mathrm{E}t) \rightarrow x = t) \rightarrow \phi)$ — 2, UG
4. $\phi^{t/x} \rightarrow (\lambda x(\Box(\mathrm{E}s \vee \mathrm{E}t) \rightarrow x = t)t \rightarrow \lambda x\phi t)$ — 3, T9
5. $\sim \phi^{t/x} \rightarrow \Box(\mathrm{E}s \vee \mathrm{E}t) \rightarrow x = t)t \rightarrow \sim \phi^{s/x})$ — 1, TF
6. $\sim \phi^{t/x} \rightarrow \forall x\Box((\mathrm{E}s \vee \mathrm{E}t) \rightarrow x = t) \rightarrow \sim \phi)$ — 5, UG
7. $\sim \phi^{t/x} \rightarrow (\lambda x\Box((\mathrm{E}s \vee \mathrm{E}t) \rightarrow x = t)t \rightarrow \lambda x \sim \phi t)$ — 6, T9
8. $\lambda x \sim \phi t \rightarrow \sim \lambda x\phi t$ — T11
9. $\lambda x\Box((\mathrm{E}s \vee \mathrm{E}t) \rightarrow x = t)t \rightarrow (\lambda x\phi t \leftrightarrow \phi^{t/x})$ — 4,7,8, TF

T13 $\vdash \forall y(\mathrm{E}y \wedge \Box(\mathrm{E}y \rightarrow \lambda x\Box((\mathrm{E}x \vee \mathrm{E}y) \rightarrow x = y)y)$

1. $\mathrm{E}t \rightarrow t = t$ — T1, Existence axiom
2. $(\mathrm{E}t \vee \mathrm{E}t) \rightarrow t = t$ — 1, TF
3. $\Box((\mathrm{E}t \vee \mathrm{E}t) \rightarrow t = t)$ — 2, Necessitation
4. $\forall y\Box((\mathrm{E}y \vee \mathrm{E}y) \rightarrow y = y)$ — 3, UG
5. $\forall y(\lambda x\Box((\mathrm{E}x \vee \mathrm{E}y) \rightarrow x = y)y \leftrightarrow \Box((\mathrm{E}y \vee \mathrm{E}y)$ $\rightarrow y = y))$ — Abstraction axiom

6.	$\forall y(\lambda x\Box((\mathrm{E}x \vee \mathrm{E}y) \rightarrow x = y)y \leftrightarrow \forall y\Box((\mathrm{E}y \vee \mathrm{E}x) \rightarrow y = y)$	5, Q distribution
7.	$\forall y(\lambda x\Box((\mathrm{E}x \vee \mathrm{E}y) \rightarrow x = y)y)$	4,6, TF
8.	$\Box\forall y(\lambda x\Box((\mathrm{E}x \vee \mathrm{E}y) \rightarrow x = y)y)$	7, Necessitation
9.	$\forall y\Box(\mathrm{E}y \rightarrow (\lambda x\Box((\mathrm{E}x \vee \mathrm{E}y) \rightarrow x = y)y))$	QCBF axiom
10.	$\forall y\mathrm{E}y$	T2, T8
11.	$\forall y(\mathrm{E}y \wedge \Box(\mathrm{E}y \rightarrow (\lambda x\Box((\mathrm{E}x \vee \mathrm{E}y) \rightarrow x = y)y))$	9, 10, Q distribution

Note that step 9 of this proof is the only application of the QCBF axiom in the proofs we have given of these theorems.

Appendix II
Completeness of First-Order S5[15]

The completeness proof is for the logic that results from combining the propositional modal logic S5 with the extensional free predicate logic with complex predicates, identity and individual constants, with one additional axiom schema, the QCBF.

The models will be variable-domain Kripke models. A model is a pair consisting of a structure $\langle \mathrm{W}, \mathrm{D}, \alpha \rangle$ and a valuation function $[\![\,]\!]$. (If ϕ is any expression, $[\![\phi]\!]$ is its *intension* [function from possible worlds to extensions], and $[\![\phi]\!]_w$ is its *extension*, relative to world w.) W is a nonempty set, the possible worlds, and D is a function assigning to each world w a set D_w of individuals—the domain of w. Because this is a free logic, the domain of any world may be empty.

The completeness proof is a construction of a model for an arbitrary consistent set of sentences. In the model we construct, W will be the nonnegative integers, with 0 being the actual world, α. The construction will result in an assignment to each world k a *saturated set* of sentences, Γ_k. We will define the domains of the worlds and the valuation function in terms of the saturated sets that correspond to the worlds. The notion of a saturated set is defined in terms of the proof-theoretic properties of the sentences, including a property of *rigidity*. Before giving the construction, I will define these concepts.

1. **Rigidity**: For any individual constant c we define a sentence RD(c) that imposes the constraint of rigidity on the name:

$$\mathrm{RD}(c) =_{\mathrm{df}} \mathrm{E}c \wedge \Box(\mathrm{E}c \supset \lambda x\Box((\mathrm{E}x \vee \mathrm{E}c) \supset x = c)c)\,.$$

[15] This long and complex proof is tailored to the particular version of first-order modal logic that I have defined, but for the most part it uses familiar methods, drawing on ideas used in completeness arguments for quantified modal logics over the years by Richmond Thomason, Kit Fine, George Hughes, and Max Cresswell, among others. The proof is specific to S5. One can prove, using a different construction but many of the same methods, completeness for quantified versions of some other systems, including K, D, K4, T, and S4.

2. **Saturated sets:** A set of closed sentences Γ of a language L is *saturated* if it is maximal consistent, and in addition, for every (open or closed) sentence ϕ, if $\exists x\phi \in \Gamma$, then for some individual constant $c, \{\lambda x\phi(c), RD(c)\} \subseteq \Gamma$.

We define L⁺ as the language L augmented by a denumerable set **R** of additional individual constants. We need the extra constants for the construction of both extensional and modal since there are consistent sets (such as $\{\sim Ft : t \in C_L\} \cup \{\exists x Fx\}$) that are not subsets of any saturated set of L. But as was shown by Leon Henkin in proofs of completeness for extensional logics, if we have a suitable supply of names that do not occur in any of the sentences of L, then any consistent set of closed sentences of original language will be a subset of a saturated set of the augmented language. The completeness proof for our version of modal predicate logic exploits Henkin's device, but it is more complicated than the proof for extensional logic in two ways: first, we need to ensure that the "witnesses" for our existential sentences use constants that satisfy the rigidity condition so that they are suitable to represent the individuals that they denote. In the semantics for some versions of quantified predicate logic, the rigidity condition is valid, but our semantics allows for individual constants that denote different individuals in different possible worlds. Second, our construction needs to build not just one saturated set of sentences (the set of all sentences that will be true in the model we define), but a set of saturated sets, each designed to be the set of all sentences that are true in one of the possible worlds of the model defined. Henkin's construction of a model for extensional predicate logic built the one saturated set in stages, starting with an arbitrary consistent set, by enumerating all the sentences of the language, and then adding the *j*th sentence at stage *j* iff it was consistent with the set obtained at the previous stage (and also adding a witness whenever the sentence at that stage had the form $\exists x\phi$). The union of all sets in this construction was a saturated superset of the original consistent set. For the modal case, we will use a similar step-by-step procedure to build a denumerable set of saturated sets at once. Instead of an enumeration of all the sentences of the language, the construction begins with an enumeration of all the pairs, $\langle k, \phi \rangle$, where k is a non-negative integer (representing a possible world) and ϕ is a sentence of the augmented language L⁺. We then define a denumerable sequence of sets of pairs $\langle k, \Gamma_j \rangle$ where k and j are non-negative integers (k representing one of the possible worlds in the model to be defined, and j representing the stage in the construction.) So for each k and j, there will be a denumerable set of pairs, $\langle k, \Gamma_j \rangle$ in this sequence. The sets that are the second terms of these pairs will be labeled Γ_{kj}. The saturated set, Γ_k, will then be defined, for each k, as the union of the sets Γ_{kj} for all j. After observing that each of these sets is saturated, we define the domains for the worlds, and the valuation function. Finally it will be shown that each sentence in the set Γ_k is true in the world k. Since the given arbitrary consistent set will be a subset of the set Γ_0 associated with the actual world of the model, this shows that any consistent set has a model.

It will be useful for the construction we will define to think of the augmented language. L^+ as built gradually, with the construction. So we define, for each *j*, the language L^j as the language that contains the extra individual constants that occur in one of the sets of sentences Γ_{kj} for some *k*. So the language L^0 is just the language L.

The inductive definition of the sets,Γ_{kj}, will be given in terms of the enumeration of pairs $\langle k,\phi\rangle$, and an arbitrary consistent set Θ of sentences of the language L. Θ is the set of sentences to be satisfied in the model we construct. For the base clause we will first define the set $\Gamma_{0,0}$, and then define all the sets $\Gamma_{k,0}$ in terms of $\Gamma_{0,0}$. For the inductive clause, the definition will also have two stages: first a specification of Γ_{kj+1} in terms of Γ_{kj}, for one of the *k*'s, and then a specification of Γ_{mj+1} in terms of Γ_{kj+1} and Γ_{mj} for the rest of the integers *m*.

The base clause: : (1) $\Gamma_{0,0} = \Theta$ (2) For each $k > 0$, $\Gamma_{k,0} = \{\phi : \Theta \vdash \Box\phi\}$. (The turnstile here means derivable in the language L^0, and more generally, when we use the turnstile in defining the sets in the construction, it will mean derivable in the language L^j.

All sets $\Gamma_{k,0}$ for $k>0$ are what I will call *necessity sets.* A set of sentences Δ is defined as a necessity set iff Δ is equivalent to $\{\Box\phi : \Delta \vdash \Box\phi\}$. That is, the deductive closure of Δ is the same as the deductive closure of the set of sentences of the form $\Box\phi$ that are entailed by Δ. For any set of sentences Δ, we will define the necessity set *of* Δ, $N(\Delta)$, as the set $\{\phi : \Delta \vdash \Box\phi\}$. Given that our modal logic is S5, this set will be a necessity set, since $\Delta \vdash \Box\phi$, iff $\Delta \vdash \Box\Box\phi$. Also since $\Delta \vdash \Diamond\phi$, iff $\Delta \vdash \Diamond\phi$, if we add any modal sentence (a sentence of either of the forms $\Box\psi$ or $\Diamond\psi$), to a necessity set, it will remain a necessity set.

The inductive clause is complicated. Here is the idea: One goes through the sequence of world-sentence pairs one at a time. If the (*j*+1)th pair is $\langle k,\phi\rangle$, then one first defines Γ_{kj+1} in terms of Γ_{kj} and ϕ, and then defines Γ_{mj+1} in terms of ϕ, Γ_{kj+1} and $\Gamma_{m,j}$ for $m \neq k$.

Specifically, the inductive clause has four parts, each specifying some additions to be made to the sets Γ_{mj} to get the sets Γ_{mj+1}. Where the (*j*+1)th pair is $\langle k,\phi\rangle$, (i) add the sentence ϕ to the set Γ_{kj} iff ϕ is consistent with that set. Otherwise, leave all the sets the same. In addition, (ii) if ϕ is a sentence of the form $\exists$F that is consistent with Γ_{kj}, also add a witness – a sentence of the form Ft, where *t* is the first individual constant in the ordered set R (the names in the vocabulary of L^+ that are not in L) that does not occur in any of the sentences added anywhere up through stage *j*. In this case one also adds the sentence RD(t) to the set Γ_{kj}. That concludes the specification of Γ_{kj+1} for this case. (iii) If ϕ is consistent with Γ_{kj} and has the form $\Diamond\psi$, then also add the sentence ψ to the set Γ_{mj} for the first *m* such that Γ_{mj} is its own necessity set. (There must be one, since at each stage of the construction there remains an infinite number of sets that are still necessity sets.) Finally, (iv) for all *m* add to Γ_{mj} the necessity set of Γ_{kj+1}. That is, for all *m*, $\Gamma_{mj+1} = \Gamma_{mj} \cup \{\phi : \Gamma_{kj+1} \vdash \Box\phi\}$.

It will be clear from this construction, that each of the sets Γ_{kj} is a subset of any of the sets Γ_{kj}, where $i > j$, and we will show that all of the sets Γ_{kj} are consistent. So the union of the sets Γ_{kj} for a fixed k and all j is a consistent set. Call this set Γ_k. These sets will be maximal, since every sentence was added at some stage if it was consistent with the set at that stage. The sets will be saturated, since the construction ensures that every existential statement has a witness for which the name satisfies the rigidity condition. Finally, at each stage j of the construction, the necessity sets, $\{\psi : \Gamma_{mj} \vdash \Box\psi\}$ will be the same for all m, since each set $\Gamma_{k,0}$ has the same necessity set, and clause (iv) ensures that this property is preserved at each stage of the construction.[16]

Here is the argument that each of the sets Γ_{mj} is consistent. First, since it is given that Θ is consistent, it is clear from the base clause of the definition that $\Gamma_{m,0}$ is consistent for all m. Second, we need to show that each of the sets Γ_{mj+1} is consistent if each of the sets Γ_{mj} is consistent. It is stipulated that the ϕ in the $(j+1)$th pair $\langle m,\phi\rangle$ is added only if it is consistent with Γ_{mj}, so that the sentence added by clause (i) retains consistency. For clause (ii): suppose for reductio that ϕ has the form $\exists x\psi$, and is consistent with Γ_{kj}, but that the witness sentence, ($\lambda x\psi t$, where t is the first unused singular term in R) is inconsistent with Γ_{kj}. That is, assume that $\Gamma_{kj} \vdash \sim \lambda x\psi t$, where t is an individual constant that does not occur in any of the sentences of Γ_{kj}, or in ψ. Then by the universal generalization rule, it follows that $\Gamma_{k,j} \vdash \forall x \sim \lambda x\psi x$, and so that contrary to supposition, Γ_{kj} is inconsistent with $\exists x\psi$. ($\exists x\psi \vdash \sim \forall x \sim \lambda x\psi x$ by the abstraction axiom.) Similarly, assume for reductio that the rigidity condition for t is inconsistent with $\Gamma_{k,j}$, so that

$$\Gamma_{kj} \vdash \sim (Et \wedge \Box(Et \rightarrow \lambda x\Box((Ex \vee Et) \rightarrow x = t)(t))$$

Since t does not occur in any of the sentences of $\Gamma_{k,j}$, it follows by UG that

$$\Gamma_{kj} \vdash \forall y \sim (Ey \wedge \Box(Ey \rightarrow \lambda x\Box((Ex \vee Ey) \rightarrow x = y)(y)), \text{ or equivalently,}$$

$$\Gamma_{kj} \vdash \forall y(Ey \rightarrow \sim\Box(Ey \supset \lambda x\Box((Ex \vee Ey) \supset x = y)(y))$$

But by T12, this implies, contrary to supposition, that $\exists x\psi$ is inconsistent with Γ_{kj}.[17] So the additions by clause (ii) preserve consistency.

[16] This last claim will be obvious for all cases where the ϕ of the $(j+1)$th pair does not have the form $\Diamond\psi$, but for that case, we need to show that when we add ψ to some set Γ_{mj}, the new set Γ_{mj+1} will have the same necessity set as Γ_{kj+1}. Here is the argument: $\Gamma_{mj+1} = \Gamma_{mj} \cup \{\psi\}$, so $\Gamma_{mj+1} \vdash \theta$ iff $\Gamma_{mj} \vdash \psi \rightarrow \theta$. Since Γ_{mj} is a necessity set, it follows that $\Gamma_{mj} \vdash \Box(\psi \rightarrow \theta)$, and so, by the S5 axioms, that $\Gamma_{mj} \vdash \Diamond\psi \rightarrow \Box\theta$. This implies that $\Gamma_{mj} \vdash \Diamond\psi \rightarrow \Box\theta$ θ iff $\Gamma_{mj} \vdash \psi \rightarrow \Box\theta$, which implies that the necessity set of Γ_{kj+1} will be the same as the necessity set of Γ_{mj+1}.

[17] Since this is a free logic that allows for sentences true in a world of a model with an empty domain (where $\forall y\phi$ will be vacuously true for all ϕ), $\forall y \sim\Box(Ey \supset \lambda x\Box((Ex \vee Ey) \supset x = y)(y))$ is not by itself inconsistent, but from it conjoined with the theorem T12, one can derive $\sim\exists x\psi$, for any ψ.

Now for clause (iii): suppose that ϕ has the form $\Diamond\psi$. Clause (iii) says that ψ should be added to Γ_{mj} for the first m such that Γ_{mj} is a necessity set, and since the necessity subsets are the same for all n, we know that $\Gamma_{mj} = \{\Box X : \Gamma_{kj} \vdash \Box X\}$. It follows from the axioms of S5, that if Δ is a necessity set, then for any ψ, ψ is consistent with Δ iff $\Diamond\psi$ is consistent with Δ, so since $\Diamond\psi$ is consistent with Γ_{kj}, ψ must be consistent with Γ_{mj}.

Now for clause (iv): This clause makes no further changes to Γ_{kj+1}, so we know that it is consistent, and so that its necessity set is consistent. Since $N(\Gamma_{kj+1})$ is consistent, we know that it is consistent with $N(\Gamma_{kj})$, and so consistent with $N(\Gamma_{mj})$ for all m. The theorems of propositional S5 ensure that for any ψ and any set Δ, $\Box\psi$ is consistent with Δ iff $\Box\psi$ is consistent with the necessity set of $N(\Delta)$, so we can conclude that $N(\Gamma_{kj+1})$ is consistent with Γ_{mj}) for all m.

The definition of the model

First, we define a partial function f taking individual constants (from any of the languages) and worlds to individual constants. $f(c,w)$ is defined as the first $d \in \mathbf{R}$ such that $RD(d) \in \Gamma_w$, and $c{=}d \in \Gamma_w$. The function is undefined for c and w such that there is no such d. It follows from the properties of saturated sets that $f(c,w)$ will be defined if and only if $c{=}c \in \Gamma_w$.

Second, define the domains as follows: For all w, $D_w = \{f(c,w) : c = c \in \Gamma_w\}$. The union of all the domains will be a subset of $\mathbf{R}$. We will call it $\mathbf{R}^*$.

Third, define the valuation function as follows:

For individual constants t, $[\![t]\!]_w = f(t,w)$. ($[\![t]\!]_w$ is undefined if $f(t,w)$ is undefined.)
For n-place predicates F, $[\![F]\!]_w = \{\langle d_1 \ldots d_n\rangle : d_i \in D_w$ for each i, and $Fd_1 \ldots d_n \in \Gamma_w\}$.
For sentence letters P, $[\![P]\!]_w = 1$ iff $P \in \Gamma_w$.

The main theorem, proved by simultaneous induction on the length of the expressions, is that for all worlds w and all assignment functions s,

1. For all closed sentences ϕ of $\mathbf{L}^+$, $[\![\phi]\!]^s_k = 1$ iff $\phi \in \Gamma_k$.
2. For all closed predicates F of $\mathbf{L}^+$, $[\![F]\!]^s_k = \{d : f(d,k) \,\&\, Fd \in \Gamma_k\}$.

These are true by definition for all primitive sentences and predicates. Most of the inductive clauses of the argument are straightforward, having three premises, one given by the semantic rule for the kind of sentence in question, one given by the hypothesis of induction, and one given by the proof-theoretic properties of the set of saturated sets. For example:

Suppose ϕ is a sentence of the form $\sim\psi$.

i.	$[\![\sim\psi]\!]^s_k = 1$ iff $[\![\psi]\!]^s_k = 0$	by the semantic rule for negation
ii.	$[\![\psi]\!]^s_k = 0$ iff $\psi \notin \Gamma_k$	by hypothesis of induction
iii.	$\psi \notin \Gamma_k$ iff $\sim\psi \notin \Gamma_k$	by the maximum consistency of Γ_k

Since the universal quantifier, in its primitive form, is a simple predicate modifier and does not bind variables, its clause is different from the more familiar one:

Suppose ϕ has the form $\forall$F (where F is a one-place predicate, simple or complex)

i. $[\![\forall F]\!]^s_k = 1$ iff $[\![F]\!]^s_k = D_k$	by the semantic rule for $\forall$
ii. $[\![F]\!]^s_k = D_k$ iff $\{d : f(d,k) = d \,\&\, Fd \in \Gamma_k\} = D_k$	by the hypothesis of induction
iii. $\{d : f(d,k) = d \,\&\, Fd \in \Gamma_k\} = D_k$ iff $\forall F \in \Gamma_k$	by the properties of saturated sets

Claim iii holds in the left-to-right direction because Γ_k is deductively closed, and in the right-to-left direction by the following argument: if $\forall F \notin \Gamma_k$, then because Γ_k is maximal, $\exists x \sim Fx \in \Gamma_k$, and because Γ_k is saturated, it must then have a witness $\lambda x{\sim}Fx(t)$. But then by consistency, ~F*t* for some *t* such that $\exists xx = t \in \Gamma_k$, from which it follows that for some $d \in D_k$, $f(d,k) = d$, but $Fd \notin \Gamma_k$.

The clause for the necessity operator is as follows:

Suppose ϕ has the form ψ.

i. $[\![\Box\psi]\!]^s_k = 1$ iff $[\![\psi]\!]^s_m = 1$	for all *m* by the semantic rule for necessity
ii. $[\![\psi]\!]^s_m = 1$ for all m iff $\psi \in \Gamma_m$	for all *m* by the hypothesis of induction
iii. $\psi \in \Gamma_m$ for all m iff $\Box\psi \in \Gamma_k$	by the properties of the construction

Claim iii holds in the right-to-left direction because as we observed above, at each stage *j*, the necessity sets of all the set Γ_{mj} are the same, which implies that $\{\Box\psi : \Box\psi \in \Gamma_k\} = \{\Box\psi : \Box\psi \in \Gamma_m\}$. So if $\Box\psi \in \Gamma_k$, $\psi \in \Gamma_m$ for all *m*. For the left-to-right direction: If $\Box\psi \notin \Gamma_k$, then $\sim\Box\psi \in \Gamma_k$, so $\Diamond \sim \psi \in \Gamma_k$. But then at the stage where the pair $\langle k, \Diamond \sim \psi\rangle$ is reached, $\sim\psi$ was added to the set $\Gamma_{j,m}$ for some *m*, and so for that *m*, $\psi \notin \Gamma_m$.

The interesting clause is the one for complex predicates. Here the argument depends on the following semantic lemma, which holds for the particular model we have defined in which the possible individuals are also names of themselves:

Semantic Lemma: For any expression ϕ (variable, individual constant, predicate, or sentence), for any $d \in \mathbf{R}^*$ and any variable x, $[\![\phi]\!]^{s[d/x]}_w = [\![\phi^{d/x})]\!]^s_w$.[18]

The members of $\mathbf{R}^*$ are both names and the objects they name; *d* plays the first role on the right side of this identity, and the second on the left side.

This lemma is proved by induction on the length of the expression. The induction steps of the argument are simple, since with one exception, both the

[18] The quantifiers in the statement of the lemma range over overlapping domains: ϕ can be any expression, including individual variables and names that are in $\mathbf{R}^*$, but the *d* must be in $\mathbf{R}^*$, and the *x* must be an individual variable.

semantic and the syntactic substitution meet the following condition: making the substitution in a complex expression is the same as making it in the immediate constituent or constituents of the expression. The one exception is the case where the complex expression is of the form $\lambda x\psi$, where the x is the same variable as the one being substituted for. In this case, the x may be free in the constituent ψ, but all occurrences of it will be bound in the complex predicate, $\lambda x\psi$. The claim is true for this case, but it requires a different argument. Here is the overall argument for the semantic lemma:

The base case: Suppose ϕ is of length one, in which case it will be a primitive descriptive symbol, or a variable. Except where ϕ is the variable x, the claim is trivial, since in these cases neither the variation in the assignment nor the syntactic substitution has any effect, so the claim reduces to $[\![\phi]\!]^s{}_w = [\![\phi)]\!]^s{}_w$.

If ϕ is the variable x, then $[\![\phi]\!]^{s[d/x]}{}_w(\phi) = d$, and the result of substituting the name d for x in x is d. So the claim holds for all expressions of length one.

The hypothesis of induction is that the claim holds for expressions of length k or less, and what we need to show is that then it holds for any expression of length k+1. First, suppose the expression is a predication sentence: a sentence of the form $\mathrm{R}t_1 \ldots t_n$. (where the t's may be either constants or variables $[\![\mathrm{R}t_1 \ldots t_n]\!]^{s[d/x]}{}_w = 1$ iff $\langle [\![t_1]\!]^{s[d/x]}{}_w, \ldots [\![t_n]\!]^{s[d/x]}{}_w \rangle \in [\![\mathrm{R}]\!]^{s[d/x]}{}_w$ by the semantic rule for predication. By hypothesis of induction, $[\![t_1]\!]^{s[d/x]}{}_w, \ldots [\![t_n]\!]^{s[d/x]}{}_w [\![\mathrm{R}]\!]^{s[d/x]}{}_w$ are equal to $[\![t_1 \, {}^d\!/\!_x]\!]^s{}_w, \ldots [\![t_n \, {}^d\!/\!_x]\!]^s{}_w [\![\mathrm{R} \, {}^d\!/\!_x]\!]^s{}_w$, respectively. Then again by the semantic rule for predication sentences, $[\![\mathrm{R} \, {}^d\!/\!_x]\!]^s{}_w = 1 \langle [\![t_n \, {}^d\!/\!_x]\!]^s{}_w, \ldots [\![t_n \, {}^d\!/\!_x]\!]^s{}_w \rangle \in [\![\mathrm{R} \, {}^d\!/\!_x]\!]^s{}_w = 1$, so the claim holds for this case. The argument for the cases where ϕ takes the form $\sim\psi, (\psi_1 \wedge \psi_2), \Box\psi, \forall\mathrm{F}$ follow the same pattern, as does the case where ϕ is a complex predicate, $\lambda y\psi$, and the variable y is different from the x. In the special case where the variable y is the same as the x that is substituted for, x does not occur free in $\lambda y\psi$ so the claim holds because neither the semantic nor the syntactic substitution has any effect on the formula or its interpretation. That is, $[\![\phi \, {}^d\!/\!_x]\!]^s{}_w = [\![\phi)]\!]^s{}_w$ (because one substitutes only for free occurrences of x) and $[\![\phi]\!]^{s[d/x]}{}_w = [\![\phi)]\!]^s{}_w$ because for any free y in ϕ, the value of $s[d, y](x) = s(x) = s(x)$. So the claim holds for this case as well. This completes the argument for the semantic lemma.

The complex predicate clause of the main argument then goes like this:

1. $[\![\lambda x\phi]\!]^s{}_w = \{d \in \mathrm{D}_w : [\![\lambda x\phi]\!]^{s[d/x]}{}_w = 1)$ by the semantic rule for complex predicates.
2. $\{d \in \mathrm{D}_w : [\![\lambda x\phi]\!]^{s[d/x]}{}_w = 1) = \{d \in \mathrm{D}_w : \mathrm{v}^s{}_w([\![\phi \, {}^d\!/\!_x)]\!]^s{}_w = 1\}$ by the semantic lemma.
3. $\{d \in \mathrm{D}_w : v^s{}_w([\![\phi \, {}^d\!/\!_x)]\!]^s{}_w = 1\} = \{d \in \mathrm{D}_w : \phi \, {}^d\!/\!_x \in \Gamma_w\}$ by the hypothesis of induction.
4. $\{d \in \mathrm{D}_w : \phi \, {}^d\!/\!_x \in \Gamma_w\} = \{d \in \mathrm{D}_w : \lambda x\phi(d) \in \Gamma_w\}$ (See below for the argument for this.)
5. Therefore, $[\![\lambda x\phi]\!]^s{}_w = \{d \in D_w : \lambda x\phi z(d) \in \Gamma_w\}$.

Argument for step 4: By saturation, $\mathrm{RD}(d) \in \Gamma_w$, so that $\lambda x\Box(\mathrm{E}x \vee \mathrm{E}d) \rightarrow x = d)d \in \Gamma_w$ for all $d \in \mathrm{D}_w$. Then by T12 $(\vdash \lambda x\Box(\mathrm{E}x \vee \mathrm{E}t) \rightarrow x = t)t \rightarrow (\lambda x\phi t \leftrightarrow \phi\, {}^{t}\!/_{x}))$ and deductive closure it follows that $\phi\, {}^{d}\!/_{x} \in \Gamma_w$ iff $\lambda x\phi d \in \Gamma_w$ for all $d \in \mathrm{D}_w$.

In conclusion, since the arbitrary consistent set Θ that we are showing to be satisfiable is a subset of Γ_0, and since all members of Γ_0 are true in the world 0 (the actual world of the model), we have shown that for any consistent set of closed sentences of the language L, there is a model in which that set is satisfied.

5

Properties and Relations

Let me sum up what I have argued so far: After setting the context for my project in Chapter 1, I began Chapter 2 by hypothesizing that there are things that are appropriately called "propositions" and then sketched a theory of them that seems to satisfy the Quinean demand that there be clear identity conditions for anything admitted to our ontology. While it was acknowledged that this regimented notion of proposition did not do all the work that propositions have been asked to do, I noted that the notion of proposition that I characterized was one to which Quine was in fact sympathetic, so long as it is recognized that it makes sense only relative to a context-dependent matrix of alternatives. I then made the seemingly innocent assumption that if there are propositions, then the sentences of a regimented language, including a language that talks *about* propositions, should themselves *express* propositions, and I used these assumptions to motivate and clarify an enrichment of the logic of the extensional language—the addition of modal operators expressing necessity and possibility. Quine was famously critical of modal logic, but I argued that his reasons for rejecting this development, while cogent against the early developers of modern modal logic, did not apply to a version of modal logic that is motivated by an ontology of propositions. The move to a modal logic did, however, raise additional questions about the role of the *predicates* of the theory of propositions, and more generally about the criteria for and consequences of the admissibility of predicates to a regimented language, a language designed to represent in a perspicuous way a theory about what things there are to talk about and what can be said about them.

Propositions. Robert Stalnaker, Oxford University Press. © Oxford University Press 2023.
DOI: 10.1093/oso/9780197647035.003.0006

I argued in Chapter 3 that the theory of propositions helps to clarify those questions and to systematize some answers to them. The intuitive idea is that a predicate, to be admissible, must bring to the table, implicitly, enough information to determine propositions when that predicate is combined, by the compositional rules of the first-order language, with quantifiers, other predicates that meet the same conditions, and names that refer. I argued that the identity conditions for propositions help to clarify a question about predicates that is analogous to Quine's general question about the identity conditions for entities admitted to one's ontology. Just as an ontological commitment presupposes that there are answers to questions about when two referring expressions refer to the same thing, so the admissibility of a predicate presupposes that there are answers to questions about when two predicates have the same theoretical role in determining propositions. It is tempting to think that when presuppositions of this kind are satisfied, that is enough to justify an ontological commitment to entities that correspond to predicates—properties and relations—but we stopped short of endorsing that commitment.

Chapter 4 was a technical digression, spelling out the details of the particular first-order quantified modal logic that I am proposing for projects of regimentation in the Quinean spirit, and of a theory of propositions formulated in a language with this logic.

So far, the regimented languages we have considered, including the language in which the theory of propositions is formalized, are all first-order languages. In this chapter, I will explore the consequences of taking the further step of accepting an ontological commitment to properties and relations, and I will consider the character of a regimented language that is appropriate for stating a theory about the relationships among individuals, propositions, properties, and relations, if such there be. So the hypothesis we will explore is that there are properties and relations that correspond to the predicates that are admissible to a regimented language.

5.1. An ontology of attributes

Quine's famous slogan about ontological commitment was that *to be is to be the value of a bound variable*. Variable binding is complex predicate formation,[1] so I have paraphrased the slogan as follows: *to be is to be a subject of predication*. This means that what we are saying if we admit properties and relations to our ontology is that these are among the things that we can say things about, which requires that we have predicates in our regimented language that apply to properties and relations. And of course the assumption that there are properties and relations expressed by the admissible predicates will apply to these higher-order predicates as well, and to the predicates with which we talk about those higher-order properties and relations. We might represent the things that our theory will be talking about with the following simple recursive definition of a categorization of the kinds of entities in our proposed ontology. First, there are two basic types: e for individuals, and a type we label with an empty pair of brackets, $\langle\rangle$ for propositions. Second, for any types $t_1, \ldots t_n$, there is a type $\langle t_1, \ldots t_n\rangle$. By "individuals," we mean whatever the ontology is of the basic theory we are generalizing. This domain could be any collection of things, since we assume that any coherent domain of entities that we are prepared to commit ourselves to can be talked about in a first-order language. The entities in any of the non-basic categories are n-ary relations that hold of an n-tuple of things of the types t_1 to t_n. For example, $\langle e, e\rangle$ is the type of a binary relation between individuals. $\langle\langle e\rangle,\langle\rangle\rangle$ is the type of binary relations between monadic properties of individuals and propositions.

This structure is exactly the type theory defined by Daniel Gallin in his semantics for higher-order modal logic.[2] It was described

[1] Implicitly, in the standard way of formulating quantification theory, and explicitly in the way I would prefer to formulate it, with a predicate-forming operator as the sole variable-binding operator.

[2] Gallin 1975. The rationale for the label for the category of propositions is that we can think of them as 0-ary relations.

there, and in later developments and applications of higher-order modal logic,[3] as a theory of types of *expressions* in the language being interpreted: names, sentences, and predicates and variables of various kinds. But it is also a categorization of the entities in an ontology of the kind we are exploring, where propositions, properties, and relations are among the things that one might talk about. The typology organizes this ontology independently of any language we might use to express, refer to, or quantify over things of these various kinds.

The typology defines *categories* of things recursively in terms of two basic categories, but it is not assumed that the membership in the derived categories can be specified as a function of the members in the more basic categories. For example, the domain of monadic properties of individuals (things of type $\langle e \rangle$) will not necessarily be determined by the relevant domain of propositions (things of type $\langle\rangle$) and the relevant domain of individuals (things of type e). But while the domains of properties and relations are not determined by the domains of lower types, there will be a rich set of structural constraints on the relationships between the different domains, and on the closure conditions for the membership for the individual categories. The main job of a general theory of propositions, properties, and relations will be to characterize the relational structure that imposes those constraints.

As discussed in previous chapters, the theorems of the basic theory of propositions include closure conditions for a domain of propositions. For example, every proposition has a contradictory, and for any set of propositions, there is a proposition equivalent to that set. There will be corresponding closure conditions on the properties and relations at each level. For example, every monadic property of individuals will have a complement, and for any set of monadic properties of individuals, there will be a property that an individual has if and only if it has all the properties in the set.

[3] For example in Williamson 2013.

Where do these conditions come from and what justifies them? The ontological hypothesis we are exploring is that a predicate should be *admissible* if and only if it expresses a property, and this means that everything we have said about admissibility conditions for predicates is relevant to judgments about the existence of properties and relations. As discussed in Chapter 3, it is a necessary condition for the admissibility of a predicate to the lexicon of a regimented language that the compositional rules of the semantics for that language should determine the propositions that sentences containing that predicate are used to express, and so the specification of the property that is its value will be the specification of a thing that can play that role. The identity conditions for the conception of property and relation we will be using will be guided by this functional role, so on the conception we are developing, properties and relations are coarse-grained in the way that propositions are, according to our theory of propositions. This is not to foreclose the possibility that there might be more fine-grained conceptions of property and relation that are useful for some purposes, just as their might be finer-grained notions of proposition that determine, but are not identified with, propositions in our sense.

We can be precise about identity conditions for properties (or more precisely, about criteria for determining when different predicates correspond to the same property), and about closure conditions on the class of properties, and this helps to clarify just what kind of thing we are taking properties and relations to be. But we cannot say much that is very precise about the general question, what is it for a property or a relation to exist? Just as we can give a description of a kind of individual and ask whether there exist any individuals of that kind, so we can describe a condition on a property or relation, and ask whether there are any properties or relations meeting that condition. The rough idea of the way such questions should be approached is this: we ask, do the facts about what the world is like provide enough information to allow us to say what a thing of the appropriate kind would have to be like to instantiate the property? As our discussions in Chapter 3 of Saul Kripke's daughter and Hillary Clinton's son brought out, it would not be enough to know, of each *actual* thing, what the world would have to be like for that thing to instantiate a property meeting a given condition. The hypothetical question we need an answer to is more general: what would the world have to be like for there to be something (perhaps something other than what actually exists) that instantiated a property meeting the condition? This is a vague intuitive question, but we have some idea how to answer it in particular cases. To use an example that Kripke discussed in his book about vacuous names,[4] we know that there are no bandersnatches, but more than that, it seems reasonable to conclude that there is no property of being a bandersnatch. The noun 'bandersnatch' comes from a fiction, and the role of names in fiction needs (and has received) special treatment, but not all non-referring names come from a fiction. Just as there are names of individual things that were introduced with the intention of referring to actual things, but that fail to do so (like the planet name 'Vulcan', discussed in previous chapters), so there are predicates introduced with the intention of describing actual phenomena, but which fail to do so. 'Being phlogiston' is a familiar example. The term 'phlogiston' is not from a fiction, but it was introduced by scientists in an attempt at reference fixing of the kind that Kripke discussed in *Naming and Necessity*. The intention was to identify a property that plays a certain role in the explanation of physical phenomena, but it turned out that there is no such property because nothing plays the right role. Because there is no such property, there is no proposition expressed by sentences that purport to use the predicate to describe things.

Of course, we do have some idea what the world would have to be like for there to be true theory that is something like the phlogiston theory just as we have some idea what the world would have to be like for the hypothesis that was presupposed by the introduction

[4] Kripke 2013.

of the name 'Vulcan' to be correct. Just as we can talk, perhaps in some detail, about a *kind* of individual, even if there is no individual of that kind, so we can talk about a kind of property that plays the role that phlogiston was thought to play, even if no property plays that role.

The closure conditions for properties of and relations between the individuals in the ontology of our basic theory will be mirrored in closure conditions on the properties of and relations between the entities of higher types. For example, if there is a monadic property X of first-order properties, then there will be a complement of that property—a property that a first-order property has just in case it lacks the property X. Our first-order logic is neutral about the character of the ontology to which it is being applied. Just as we can have a first-order theory whose domain is a set of propositions, so we could have a first-order theory of the properties and relations of the members of some other domain. But our aim will be to construct a theory that is suitable for talking about the relations between all the properties and relations of the different types in the hierarchy, a theory that can describe the closure conditions on the membership of the individual type-categories and the conditions that the membership of one category imposes on the membership of others.

A Quinean might want a theory that described this structure in a first-order language, with the entire ontological hierarchy constituting the domain of the quantifier,[5] or perhaps a first-order language with separate variables and stratified quantifiers, each

[5] Though keep in mind that we are not assuming that our regimented theory is a comprehensive theory of absolutely everything that exists, or even that the idea of such a comprehensive theory makes sense. So a domain that included "the entire ontological hierarchy" means a domain built by extending a domain of individuals that is the subject matter of the given regimented first-order theory we start with. When we first proposed to explore an ontology of propositions, we acknowledged that we could perhaps make clear sense of a domain of propositions only in a particular context, and that the maximal propositions of the theory, in application, might make just the distinctions that are made in that context. This was a concession to Quine, who allowed that one might make sense of propositions in limited contexts, but it is an acknowledgment I think we should make anyway.

ranging over the domain of entities of each type. (The kind of language we considered at the end of Chapter 4 for a theory about a certain domain of individuals, plus propositions about them.) We will consider the first-order option at the end of this chapter, but we will here follow the standard practice, using a higher-order modal language—a type theory of the kind developed by Gallin and used by Williamson and others—to describe this relational structure. Quine would warn us that by choosing this path we are dressing our theory of properties in sheep's clothing, blurring distinctions in a way that makes it easier to equivocate. I think Quine's warning should be taken seriously as a caution, but I also think that with care we can avoid equivocation. Before discussing the higher-order modal logic, I will digress by saying what the potential confusions are, according to Quine, that might be involved in allowing a logic that quantifies with variables for predicates.

5.2. Property theory in sheep's clothing

Quine's famous quip that second-order (extensional) logic was "set theory in sheep's clothing" occurs in his *Philosophy of Logic* in a discussion of the scope of logic.[6] While at the end of the day, he was prepared to admit sets into his ontology, he rejected the view that set theory was a part of logic. "The tendency to see set theory as logic has depended early and late on overestimating the kinship between membership and predication." A simple monadic atomic sentence involves a predicate and a name. In any set-theoretic model for the language, there will be a set associated with the

[6] The quip has a point, but the idea that second-order logic is a version of set theory can't be taken literally. See Boolos 1975, who points out that there are (set-theoretic) models of second-order logic where there are only sets with no more than one member in the domain of the second-order quantifier. This paper precedes Boolos's important work developing a plural interpretation of second-order logic, which he argues is not committed to the existence of sets at all.

predicate—its extension—and the sentence will be true if and only if the referent of the name is a member of the set that is associated with the predicate. But the predicate does not *name* the set that is its extension, and Quine argued that it is important to distinguish name position in a sentence from predicate position. One cannot explain predication in terms of the set that is the extension of the predicate since we need another predicate (the binary membership predicate) to state that a thing is in the extension. That is, in the set-theoretic statement '$t \in \text{Ext}(F)$', it is the binary membership symbol that is in predicate position. The distinction between name position and predicate position, Quine observed, is important also for those who are "fond of attributes." They "should not read 'F*x*' as '*x* has F', with 'F' in name position [referring to an attribute]; rather let him write '*x* has *y*', or if he prefers distinctive variables for attributes, '*x* has ς.'"

As Quine is using the terms, 'name position' and 'predicate position' refer to the role of an expression in predication sentences—sentences of the form $Ra_1, \ldots a_n$, where R is an *n*-place predicate and $a_1 \ldots a_n$ are expressions whose values are entities to which the relational predicate R is applied. R is in predicate position, while $a_1 \ldots a_n$ are in name position.[7] In the higher-order languages, R might be a predicate of type $\langle t_1 \ldots t_n \rangle$ and $a_1 \ldots a_n$ of types $t_1 \ldots t_n$, respectively, for any *n* types. The problem with higher-order languages, according to Quine, is that they allow the same expressions to occur in both name and predicate position, and more seriously, when one uses predicate variables to quantify over the entities that are the semantic values of predicates, one treats a variable in predicate position as if it were in name position: "To put the predicate

[7] Nick Jones thinks that the terminology "name position" is misleading and prefers the term "argument position," restricting the former term to expressions of type *e*. He suggested (personal communication) that Quine's objections to higher-order logics would be defused if we used his preferred terminology. I agree that Jones's preferred label is clearer, but I don't see why the terminological issue is relevant to Quine's concerns, or why a higher-order language could not have names for things of other types than *e*.

letter 'F' in a quantifier, then, is to treat predicate positions suddenly as name positions, and hence to treat predicates as names of entities of some sort." Quine thought that use-mention conflation motivated some logicians who made this move (they implicitly take the second-order quantifier to be quantifying over predicates rather than things that are the values of predicates), and I think some defenses of quantification with variables in predicate position lend support to this conclusion. I will point to one example.

Use/mention conflation in the defense of higher-order quantification is explicit in a chapter of Arthur Prior's posthumously published book, *Objects of Thought*.[8] It is not that Prior confused expressions with what they designate—he carefully distinguished them—but he then blurs the line in the interpretation of the variables in open sentences and in the interpretation of quantification. Prior began by asking what a variable in an open sentence *stands for* and then distinguished two interpretations of what we might mean by 'stands for' in this context. "The variable may be said, in the first place, to stand for a name (or to keep a place for a name) in the sense that we obtain an ordinary closed sentence by replacing it by a name, i.e. by *any* genuine name of an individual object or person." Alternatively, we might use 'stands for' in a secondary sense in which it is the referent of a name—the individual object or person named, presumably by a name that the variable 'stands for' in the first (more basic) sense. Now when we come to a predicate variable in an open sentence such as 'ϕ' in 'Peter ϕ 's Paul', "it is easy to say what 'ϕ' or 'ϕ's' ... 'stands for' in the first sense—it keeps place for any transitive verb, or any expression that does the job of a transitive verb," meaning any binary predicate. But the "question what it 'stands for' in the second sense, i.e. what would be designated by an expression of the sort for which it keeps a place, is senseless, since the sort of expression for which it keeps a place is one which just hasn't the job of designating objects."

[8] Prior 1971.

These remarks might be innocent if we are just interpreting open sentences that remain open, but what are we doing when we bind one of these higher-order variables? When we bind a variable that "keeps a place" for a *name*, we interpret the result in terms of a domain of entities of the sort that names 'stand for' in the secondary sense, but how does Prior interpret a quantified statement, where the variable that is bound is of the sort that hasn't the job of designating objects? No problem, he suggests: " 'For some ϕ, Peter ϕ 's' is true if any specification of it is true, meaning by a 'specification' of it, any statement in which the indefinite verb [the variable]... is replaced by some specific verb or equivalent expression." I don't see any way to understand this except as an endorsement of a substitutional interpretation of the quantifiers. Perhaps one can give a coherent account of this kind of interpretation, though more needs to be said about the range of possible substitutions. It is not, however (contrary to what Prior says), a simple generalization of objectual quantification—what Prior calls "nominal quantification." It is a familiar point that there are many unnamed things in many domains that an objectual quantifier ranges over, for example most real numbers, and most grains of sand. The idea of a domain of entities to be referred to or quantified over is not, as Prior suggests, derivative from the idea of a domain of names that such things might have. Taking the referential sense of "stands for" to be a secondary sense is the basis for saying that the explanation for quantification into sentence and predicate place is a straightforward generalization of first-order quantification. ("All this carries over, mutatis mutandis, into the discussion of quantifications over variables of other categories.") But if a proper explanation of nominal or objectual quantifiers is to carry over to the higher-order case, we need the hypothesis that there is a domain of entities of some kind in order to understand what those quantified sentences are saying. If one agrees with Quine, as Prior seems to, that there are no entities (properties and relations) corresponding to predicates, then one should agree with Quine in rejecting higher-order quantification rather than dismissing the meta-logical thesis that quantification presupposes an ontological commitment as "a piece of unsupported dogma."[9]

Williamson, who seems to agree with Prior that quantification into predicate position should not be understood as quantification over a range of objects, recognizes that a substitutional interpretation of the quantifiers is not appropriate for the kind of higher-order quantification that he endorses, and he also observes that the plural interpretation of second-order quantifiers pioneered by George Boolos is not the right way to understand a higher-order language that not only permits quantification into predicate position but also allows predicates that apply to the values of the predicate variables.[10] So how, according to Williamson, are we to understand the kind of quantification that is allowed in higher-order modal logic, and the application of predicates to higher-order entities that are not among the things included in the domain of absolutely everything? Here is what he has to say about this question:

> Perhaps no reading in a natural language of quantification into predicate position is wholly satisfactory. If so, that does not show that something is wrong with quantification into predicate position, for it may reflect an expressive inadequacy in natural languages. We may have to learn second-order languages by the direct method, not by translating them into a language with which we are already familiar. After all, that may well be how we come to understand other symbols in contemporary logic.... We must learn to use higher-order languages as our home language.

[9] Quotations in this paragraph are from Prior 1971, 34, 35, and 48.

[10] Plural quantification allows for quantifiers over all the things, or some of the things that are F without presupposing that there is a single entity (a set or some other kind of collective object) that is a plurality of things that are F. To apply a predicates of a regimented plural language to the Fs, or to some of the Fs, is to apply that predicate to each of the relevant Fs. It is a further step that one might think makes an ontological commitment to plural objects of some kind to allow for plurals that describe collective action, as when one says that the movers carried the piano up the stairs or that the soldiers surrounded the town. See chapters 4 and 5 of Boolos 1998b.

I agree that we are not looking for a synonymous translation of the notation of higher-order language into natural language. Regimentation involves revision and refinement of the resources provided by the language we begin with. But I don't see a problem with a naïve understanding of the higher-order quantifiers as ranging over a domain of entities that we can describe with the still higher-order predicates of the language. I don't see a problem with the following description, in natural language, of how to understand what this regimented language (higher-order modal logic) is about: There is a range of entities of different categories, starting with a basic category of individuals which could be any collection of things we think there is in the world to talk about), and a basic range of admissible predicates for describing those things. The descriptions of the individuals express propositions—another basic category of things—and then there are further entities corresponding to the predicates, and a range of further predicates for describing those further things, and for describing some relations between the things of the various categories, and then further objects corresponding in the same way to the further predicates, and so on. What I don't understand is how to reconcile this way of understanding the higher-order language with the claim that there is in this language an objectual quantifier that ranges over absolutely everything, with the claim that there are other quantifiers in this language that range over some additional things. It doesn't help me to be told that once I have adopted the higher-order language as my home language, I will see that my naïve understanding of what I am doing with that language is incorrect.

I think Quine is right that the blurring of the line between expressions and their values plays a role in some explanations of higher-order quantification, but he acknowledged that this is not essential. "The logician who grasps this point [that a quantifier (∀F), where F is a predicate variable, is quantifying, not over predicates, but over entities that are the values of predicates], and still quantifies 'F', may say that these entities are attributes; attributes are, for him, the values of 'F', the things over which 'F' ranges." This is perhaps okay, if one is prepared to make and to defend an ontological commitment to attributes of all the appropriate types, just as second-order extensional predicate logic is perfectly intelligible to one who accepts the existence of sets. But one should resist the idea, apparently endorsed in Prior's and Williamson's defenses of quantification into predicate position, that one is making no further ontological commitments by enriching one's logic in this way.

Quine's distinction between predicate and name or argument position remains important on the Quinean interpretation of higher-order modal logic, and if we are going to be "the unconfused but prodigal logicians with our eyes open," we need to recognize that we are mixing the two roles by allowing predicates to occupy them both. In the higher-order language with predicate quantifiers, a first-order predicate F may be in predicate position as in a sentence of the form F*t* (where *t* is a name of an individual), or in name position, as in the sentence G(F), where G is a second-order predicate, expressing a property of properties of individuals. A predicate *variable* may play both roles, as in a sentence of the form $\exists f(Gf \wedge ft)$, which says (roughly) that there exists a property *f* of individuals such that *f* has G, and *t* has *f*. We blur the distinction Quine took pains to make between F*x* (where F *expresses* the property ζ) and *x* has ζ (where ζ *names* the property that is *expressed* by F, and the predicate ('has') expresses the exemplification relation between an individual and a property).[11]

So we are still following Quine in granting that a higher-order language that mixes roles in this way is intelligible, given the ontological commitment to properties and relations corresponding

[11] Our language may have predicates for the binary exemplification relations, and so, for example, sentences of the form E(F,G), where E expresses the exemplification relation, F denotes a monadic property of individuals, and G denotes a monadic property of properties. Given our coarse-grained conception of proposition, there will be a close correspondence between 'E(F,G)' and 'G(F)', but they will not necessarily express the same proposition.

to the predicates, and to propositions expressed by the sentences. We can gain considerable simplification in the task of articulating a theory that describes the structure of relations between the members of the different categories in our ontological hierarchy if we blur the line between predicate and name position in this way. What is the cost? The general concern underlying Quine's qualms about higher-order logic is the need to be clear about the relation between issues about the language with which we talk about things and issues about the things themselves. It is relatively easy to avoid confusing names of particular physical objects with the objects themselves, but when we talk, in a semantic meta-language, about the roles of predicates, there are new distinctions that complicate the issues. The hypothesis that there are attributes is the hypothesis that just as the semantic role of a name is explained (at least in part) in terms of a relation between the name and the individual, so the semantic role of a predicate (even in the semantics for a first-order language) is to be explained (in part) in terms of its relation to an attribute, and given this hypothesis, one might label the relevant attribute as the *semantic value* of the predicate. But specifying the entity that is the semantic value of a predicate in this sense does not suffice to characterize its semantic role. Entities of any kind can be named, and the role of a name in a compositional semantics is different from the predication role. Both roles are to be explained (in a semantic meta-language) in terms of the way the expression contributes to the complex expressions of which it is a constituent. In a language with variables and variable-binding operators, the importance of distinguishing semantic value and compositional role is particularly important, and the distinction remains important even if we allow predicates and predicate variables to play the two roles at once.[12]

[12] I think a certain amount of unnecessary puzzlement about semantic values can be explained by a tendency to presuppose that the semantic value of an expression must be both an entity of some kind, and something that determines the compositional role of the expression. Given this presupposition, it seems that the Fregean sense of a predicate

Predicate position and name position are not the only compositional roles. There are also *sentential* and *operator* roles. The semantic value of a sentence is a proposition or, in an extensional theory, a truth-value. Sentences *express* propositions, but in our first-order theory of propositions, there might also be names for them (as the names, 'η', 'κ', and 'α' that we introduced for the necessary, the impossible, and the true maximal consistent propositions). In both first-order and higher-order languages, there are operators and connectives, and one may take their semantic values to be functions taking propositions, or pairs of propositions, to propositions. But even though functions are relations—entities that may be the values of predicates—operators and connectives are not predicates. But blurring the lines between roles makes it easier to ignore the distinction between propositional operators and predicates of propositions, and so easier to equivocate between them. If we allow sentences to occur indifferently as expressing and as naming propositions, then sentential operators will look a lot like predicates of propositions. If *o* is a sentential operator and G is a predicate of propositions, then for any sentence ϕ, both *o* ϕ and G ϕ will express (and denote) propositions that are a function of the proposition expressed by ϕ, but in the former case the ϕ has the expressing role, while in the latter case the ϕ has the naming role.

As was emphasized in Chapter 3, a central role of predication is to determine a proposition as a function of the entity to which the predicate is ascribed. It follows that a central role of a predicate *of propositions* is to determine a proposition as a function of a proposition. But we can't assume that the application of any function from propositions to propositions is therefore a case of predication—a case of saying something about the argument of the function. Predication and ontology are interdependent concepts, since

can't be an entity that one might name and talk about, but must remain elusively unsaturated. Perhaps one can make sense of Frege's remarks about the mysterious concept horse, but I think they should be interpreted as an attempt to solve an illusory problem. Similarly for Wittgenstein's remarks about what can be shown, but cannot be said.

ontological commitment is a commitment to the existence of a domain of things that are described by a certain range of predicates. What Williamson calls 'the being constraint' is a consequence of this way of understanding predication: A proposition that ascribes a predicate entails the existence of the entity to which the predicate is ascribed. In the context of the Quinean methodology that we are presupposing (extended as we have proposed to extend it), to question this constraint would be to question the intelligibility of the idea of ontological commitment.

As we observed in the discussion of Quine's three grades of modal involvement in Chapter 2, the predicates of the theory of propositions give us the resources to specify the semantics for a modal operator that extends our original language. The semantic rule for any singular operator '**o**' must say what proposition is expressed by a sentence of the form **o**ϕ as a function of the proposition expressed by ϕ. A function is a special case of a binary relation (a binary relation R is a (total) function if and only if for every x in the domain, there is a unique y such that Rxy), so the semantics for any operator will specify a binary relation between propositions that is definable in the theory of propositions. For example, we defined the relation *being the contradictory of* in terms of the primitive predicates of the theory of propositions, and it is a theorem of the theory that this relation is a function. This function provides the semantic rule for the negation operator: any sentence of the form $\sim\phi$ will expresses the proposition that stands in this relation to the proposition expressed by ϕ. But the sentence $\sim\phi$ is a first-order sentence; it does not make a statement *about* the proposition expressed by ϕ.

Given the theory of propositions and the presupposition that the sentences of our regimented language express propositions, we can specify, for any operator that has a well-defined semantics of this kind, a monadic predicate of propositions that corresponds to the operator. The correspondence between operator and predicate, in general, will be this (using the language that blurs the line between the expressing role and the naming role): $\forall p(Op \leftrightarrow \mathbf{o}p)$ (where **o** is the operator, and O is a predicate expressing the corresponding property of propositions), and this correspondence will be necessary, so that $\Box\forall p(Op \leftrightarrow \mathbf{o}p)$. In the case of negation, the corresponding property is the property of being the contradictory of a true proposition—in other words, the property of being a false proposition. In the case of the possibility operator, the corresponding property is consistency. But it would be a mistake to take the operator to *be* a predicate, and a mistake to assume that O*p* must express the same proposition (for a given value of *p*) as **o***p* does for that value. The potential difference between the two propositions is this: An atomic predication sentence (of any type) entails the existence of the thing named. Suppose, as some philosophers believe, that (at least some) singular propositions about contingent things are themselves contingent things that depend for their existence on the existence of the thing the proposition is about. (Call this "the object-dependence assumption.") So we are supposing, for example, that where P is the proposition *that Hillary Clinton has a son*, the proposition *that P exists* entails the proposition *that Hillary Clinton exists*. If we assume further that the proposition *that Hillary Clinton exists* is a contingent proposition, we can derive the conclusion that there is a proposition *p* such that possibly both ~*p* and ~F*p*, (where 'F' is the predicate of falsity), which implies that in this case the proposition expressed by 'F*p*' cannot be the same as the one expressed by ~*p*'.

Here is the argument, where 'P' expresses the proposition *that Hillary Clinton has a son*, 'H' expresses the proposition *that Hillary Clinton exists*, 'F' is the falsity predicate, and 'E' the existence predicate. The four premises are as follows:

(1) $\Box(\mathrm{EP} \rightarrow \mathrm{H})$ Object dependence
(2) $\Box(\mathrm{FP} \rightarrow \mathrm{EP})$ The being constraint
(3) $\Box(\mathrm{P} \rightarrow \mathrm{H})$ The being constraint
(4) $\Box{\sim}\mathrm{H}$ Contingency assumption

(5) $\Box(\mathbf{F}\mathrm{P} \rightarrow \mathrm{H})$ from (1) and (2)
(6) $\Box(\sim\mathrm{H} \rightarrow (\sim\mathrm{P} \wedge \sim\mathbf{F}\mathrm{P}))$ from (3) and (5)
(7) $\Box(\sim\mathrm{P} \wedge \sim\mathbf{F}\mathrm{P})$ from (4) and (6)
(8) $\exists p\Box(\sim p \wedge \sim \mathbf{F}p)$ from (7) by existential generalization (presupposing the existence of proposition P)

The conclusion is equivalent to the negation of $\forall p\Box(\sim p \rightarrow \mathbf{F}p)$, so since the thesis, $\forall p\Box(\sim p \leftrightarrow \mathbf{F}p)$ is true, the conclusion provides a counterexample to the converse Barcan principle.

The identity function from propositions to propositions corresponds in this way to the truth predicate, so the following will be true: $\forall p\Box(\sim p \leftrightarrow \mathbf{T}p)$, but for the same reason,[13] we cannot conclude that the proposition that a certain proposition is true will be the proposition itself.

If one conflates the operator with its corresponding predicate, one might be tempted to argue like this: Any tautology will express the necessary proposition, so for any proposition p, $\Box(p \vee \sim p)$. But p is the same as $\mathbf{T}p$, and $\sim p$ is the same as $\mathbf{F}p$, so we can conclude $\Box(\mathbf{T}p \vee \mathbf{F}p)$. But by the being constraint, both Tp and Fp entail that p exists, so we can conclude $\forall p\Box\mathbf{E}p$ – all propositions exist necessarily. Some philosophers—necessitists—are happy with this conclusion, and more generally the conclusion that everything (in any of the type-categories) exists necessarily, but they need a necessitist premise to justify the identification of the proposition expressed by the operator sentence with the proposition expressed by the corresponding predication.

[13] For an argument that there can be a proposition p such that $\Diamond(p \wedge \sim \mathbf{T}p)$, just substitute, in the argument given above, ~p' for 'P' in premises (1) and (2), but leave premises (3) and (4) as they are. Replace the predicate 'F' everywhere with 'T'. The substitutes for lines (5), (6), and (7) and conclusion (8) should read as follows:

(5*) $\Box(\mathbf{T}\sim p \rightarrow \mathrm{H})$
(6*) $\Box(\sim\mathrm{H} \rightarrow (\sim\mathrm{P} \wedge \sim\mathbf{T}\sim\mathrm{P}))$
(7*) $\Diamond(\sim\mathrm{P} \wedge \sim\mathbf{T}\sim\mathrm{P})$
(8*) $\exists p\Diamond(p \wedge \sim\mathbf{T}p)$

Because the higher-order language allows variable-binding into positions other than name position, lambda abstraction makes it possible to, in a sense, define predicates of propositions in terms of the corresponding operators[14] in the following way: where p is a propositional variable, $\lambda p \sim p$ is the falsity predicate, and $\lambda p \Diamond p$ is the consistency predicate (and λpp is the truth predicate). If lambda conversion were valid, in general, one could go back and forth between $\sim p$ and Fp, and between $\Diamond p$ and Cp (and p and Tp). But in our first-order semantics, lambda abstraction and its converse are valid only within the scope of a quantifier, and this feature will carry over to the higher-order logic, so these inferences will be fallacious. But variable binding into sentential position encourages the fallacy.

5.3. Properties and propositional functions

Peter Fritz, who explicitly endorses the identification of sentential operators with propositional predicates, accepts the validity of an argument I have judged to involve equivocation, but he suggests that the contingentist should apply modus tollens to its conclusion, rejecting the being constraint, the claim that predication entails existence.[15] (He does accept a weaker constraint: that predication

[14] I say "in a sense" because these definitions do not provide any kind of reductive explanation of the predicates. My reason for saying this is that the semantic rules for the operators are specified with the predicates of the theory of propositions, so it would be circular to take the complex predicate defined in terms of the operators as a reductive definition. This points to one of the ways that higher-order logic may be misleading. The higher-order logic presupposes a prior commitment to propositions, which is a commitment to the admissibility of the primitive predicates (truth and consistency) of the first-order theory of propositions.

[15] Fritz 2020. Fritz's argument assumes that the material conditional connective is a binary predicate, expressing a binary relation between propositions. Given this assumption, a sentence of the form $p \rightarrow p$ has the logical form Rpp, and so the sentence $((p \rightarrow p) \rightarrow \mathrm{E}p)$ is an instance of a schema that seems to be valid, given the existence axiom. If we assume the necessitation and generalization of instances of this schema, we can infer $\Box\forall p(p \rightarrow p) \rightarrow \mathrm{E}p)$. Since $(p \rightarrow p)$ is a tautology, $\Box\forall p\Box(p \rightarrow p)$ will be valid, and so it will follow that $\Box\forall p\Box\mathrm{E}p$ – all propositions necessarily exist. The reasoning is unproblematic. The problem, I think, is the assumption that connectives are predicates.

entails *possible* existence.) I find the denial of the being constraint unintelligible. I can't understand the claim that merely possible things have properties and stand in relations as a claim that says something different from the claim that there exist merely possible things. Of course I do understand how there could be predication sentences that are in fact not true, but that *would* be true if things were different from the way they are, and while I don't think it is intelligible to say that things can have properties and stand in relations even when they don't exist, I do, however, see (dimly at least) how it might be intelligible to deny the validity of a logical principle $\Box(Rt_1, \ldots R_n \rightarrow Et_i)$, that is an axiom of our first-order modal logic. The thesis that this principle is valid is not the being constraint itself, but a principle that is motivated by a certain intuitive conception of what predication is, and of what it is for a predicate to be admissible in a regimented language. On this conception of predication, explored and defended in Chapter 3, the role of a predicate is to distinguish between the members of some given domain of entities. For a predicate to be admissible is for the world to provide the resources for saying, about any possible situation and arbitrary individual of the appropriate kind that exists in that situation, whether the predicate is true of it (or in the general case, whether the predicate is true of an *n*-tuple of individuals of the appropriate kind that exist in that possible situation). This is the rough intuitive characterization of what it is for a predicate to be admissible. To clarify this idea in a formal possible-worlds model, the semantic value of a predicate F is a rule that picks out some of the things in the domain of the possible world—those that are the Fs. Call this the property view of predication. I think the conception is well motivated, and it is neutral on the metaphysical questions about necessitism, and it validates the existence axiom. But there is an alternative way to understand predicates which I think is coherent (though it raises some questions I am not sure how to answer) and that does not validate the existence axiom. Call this alternative conception the propositional function view: Predicates, on this conception, should be understood as expressing functions from individuals to propositions.

A function, understood extensionally, is defined for a certain domain of arguments and range of values. For propositional functions, the domain consists of the individuals. If functions f and g take each individual in the domain to the same proposition, then $f = g$. In the discussion of predicates and predication in Chapter 3, I argued that it was an essential feature of an admissible predicate that it should *determine* a propositional function, but one of the central arguments of that chapter was that (given certain perhaps contentious, but coherent metaphysical assumptions) there were predicates that play different compositional roles in determining propositions, but that corresponded to the same propositional function. This is possible since the role of a predicate is not only to determine a singular proposition for each member of the domain but also to determine the proposition expressed by quantified statements that are true if and only if the predicate applied to all (or to some) of the things in the relevant domain. The argument was given in the informal pre-regimented language and was based on the judgment that certain predicates were admissible, and on pre-regimented judgments about the result of applying those predicates to individuals in counterfactual situations.

The examples, recall, were the predicates 'is a daughter of SK' and 'is a son of HC'. The metaphysical assumption was that there is nothing in the relevant domain of individuals that could have been either SK's daughter or HC's son. This is contentious since the necessitist holds that there actually exist things that could have had these properties, even though these entities are not human beings and are not located in space and time. On the necessitist metaphysics, human beings are not essentially human or essentially located in space and time. The logical framework I have defended is compatible with the necessitist metaphysics but also compatible with its denial.

The conception of property and relation that I am defending does not provide any kind of reductive analysis of the entities we are supposing to exist. What is assumed is, first, that if we judge that a predicate is admissible, then we should judge that it expresses a property or relation, and, second, that if two predicates have different compositional roles in determining propositions, then they express different properties. So one upshot of the argument is that if we want to allow for the possibility that there might have been things that do not exist, we should not identify properties with (extensional) propositional functions.

We might, however, understand a function intensionally as a rule that could be applied in counterfactual situations, a rule that determines an extensional propositional function for that counterfactual situation. That is, the rule would take each individual existing in the counterfactual situation to a proposition that exists in that situation. An intensional propositional function of this kind does not violate the being constraint since the rule is applied in a counterfactual situation and is defined only for arguments and values that exist in that situation. It does, however, invalidate the existence axiom since the value of a propositional function, for argument *d*, might be a proposition that does not entail the existence of *d*. (This interpretation of the propositional function view should be distinguished from the idea of simply adding merely possible things to the domain of the extensional function, as applied in the actual situation. That would be the proposal that I find unintelligible, since one cannot apply a rule to a thing that does not exist.)

On the intensional propositional function view, the values of predicates contain enough information to distinguish the compositional role of the predicate 'is a daughter of SK' from that of 'is a son of HC', but there are further problems. Consider a possible situation in which HC not only has no sons but does not herself exist. In such a situation (according to the contingentist), there would be no propositions of the form that x is a son of HC, and so even though 'x is a son of HC' determines a proposition for each value of x in the domain of the actual world, there will no proposition to be a value for the propositional function in the counterfactual situation.[16] Even though the property of being a son of HC would not exist if that counterfactual possible situation were realized, the predicate 'is a son of HC' is admissible in the actual world, and (as we argued in Chapter 3) it seems to be *applicable* in the counterfactual situation. It would be false of everything, since nothing would be HC's son if she did not exist. The predicate 'is not a son of HC' would apply to everything in such a situation. This is what the property view says.

If we assume necessitism—the metaphysical thesis that it is necessary that everything necessarily exists—then the two alternative ways of understanding predication will be essentially equivalent: there will be no cases where the extensional propositional function determined by a predicate is insufficient to determine a property. But for the more general case, where the domains of both individuals and propositions may be different in different possible situations, they come apart. There may be ways of developing the propositional function view, perhaps combined with an explanation of how to understand quantifiers that range over things (or at least *cases*) that are outside of the domain of absolutely everything,[17] so that it yields an adequate representation of a contingentist metaphysics, but even if there is, I doubt that the propositional function conception of predication yields a non-question-begging argument for necessitism. But one needs to look more carefully and in more detail at the propositional function view and at interpretations of the type theory that are different from my Quinean interpretation of it. This is further work to be done.[18]

[16] Bruno Jacinto pointed this out to me several years ago in correspondence about the problem of reconciling contingentism with model-theoretic semantics for modal logic.

[17] Fritz, like Prior and Williamson, wants to allow for quantification without ontological commitment. He argues that if contingentism is true, there are *cases* of mere possibilia, even if there exist no *things* that are merely possible. I would rather say that there are *properties* of being a thing of a certain kind that are in fact uninstantiated.

[18] For discussion and development of such interpretations, see (in addition to Prior 1971, Fritz 2020, and Williamson 2003) J. Goodman 2016, Dorr 2016, Bacon 2018, Jones 2018, Florio and Jones 2012, and Rayo 2020. Thanks to Jeremy Goodman and Agustin Rayo for helping me get at least a limited grasp of these ideas.

5.4. A theory of propositions, properties, and relations

The upshot of our consideration of operators and predicates, properties and propositional functions, is this: Stating a theory of propositions, properties, and relations in a language that blurs the line between the expressing and the naming role is not by itself a confusion, and need not involve equivocation, but it does make equivocation easier, and it can make fallacious reasoning look more plausible. It is okay to dress our theory in sheep's clothing so long as we are careful to distinguish the wolves from the sheep that they are dressed to resemble. The advantage we gain by formulating the theory in this kind of logical framework is that we can more easily describe the structure of relations among propositions, properties, and relations by representing them by the ways they could be expressed with the compositional resources of the language. Let me illustrate some of the principles that describe that structure.

One of the constraints on the relation between attributes that we discussed informally earlier is that for every binary relation and individual, there is a monadic property that something has if and only if it stands in the binary relation to the individual (recall, again, the property of being Hillary Clinton's son.) We could state this particular closure condition in the higher-order language with the following sentence, abbreviated as X:

$$\mathrm{X} =_{df} \forall r \forall x \exists f \Box \forall y (fy \leftrightarrow rxy)$$

r is a variable of type $\langle e, e \rangle$, f is a variable of type $\langle e \rangle$, x and y are variables of type e. Thesis X is a *pure* sentence, generalizing about properties and relations, but containing no predicates, and no reference to any particular individual, property, or relation. Like all of the conditions describing the necessary structure of the ontological hierarchy (closure conditions on the domains of different types, and constraints that the domains of propositions and attributes of different types impose on each other), it does not make categorical claims about what propositions and attributes exist, but only conditional claims about what propositions and attributes there must be, given that there are certain others. There will be some necessarily existing properties and relations that are essential to there being individuals, properties, relations, and propositions at all (most prominently, the identity relation and the properties expressed by the primitive predicates of our theory of propositions), but the domains of empirical properties and relations, like the domain of empirical individuals, are provided for us by the world as it happens to be. Logic and the general theory of properties and relations should leave open whether the entities (of all types) that the world provides us are things that exist necessarily.

Thesis X generalizes: r might be a variable of any type $\langle t_1, t_2 \rangle$, f a variable of type $\langle t_2 \rangle$, x a variable of type t_1, and y a variable of type t_2. Still more general: r might be an n-ary relational variable for any n, and f an appropriately typed $(n-1)$-ary relational variable, and so on. There will be other comprehension principles, or closure conditions, of the kind we have discussed corresponding to the logical operators and connectives in terms of which predicates are defined. Using them, one will be able to prove a general comprehension schema of the following form:

Let r be any variable of type $\langle t_1, \ldots t_n \rangle$, and let $v_1, \ldots v_n$ be variables of type $t_1, \ldots t_n$, respectively. Let ϕ be any pure sentence (i.e., sentence with only logical symbols and variables) that does not contain r. Then let M* be the following sentence:

> $\exists r \Box \forall v_1 \ldots \forall v_n (r v_1 \ldots v_n \leftrightarrow \phi)$. Let 'U' abbreviate a string of universal quantifiers binding all of the free variables in ϕ.

Finally, let C* be defined as UM*. Since this general comprehension schema is valid, so is its necessitation.

Like X, instances of the schema C* are all pure sentences, containing no names or predicate constants. To apply them to get conclusions about the existence of properties and relations corresponding to particular defined complex predicates, we need existential premises about the primitive predicate constants in terms of which the complex predicates are defined. Just as, in our free first-order logic, one needs a premise, $\exists x x = t$ to infer Ft from $\forall x$Fx, so in a higher-order logic one needs existential premises stating that there exists a property or relation expressed by the relevant primitive predicates. So, for example, if f is a monadic predicate variable of type $\langle e \rangle$ and G is a predicate constant of that type, then $\exists f \Box \forall x(fx \leftrightarrow \mathrm{G}x)$' will say, in effect, that there exists a property expressed by 'G'. While every instance of the schema C* is a necessary truth, the existential premises needed to instantiate these universal generalizations may be contingent truths.

Applied sentences that assert the existence of certain properties expressed by complex predicates might still be, in a sense, logical truths, even if they are contingent truths. Recall the notion of *real-world validity* defined by Bruno Jacinto and discussed in section 4.3 of Chapter 4. Real-world validity was contrasted with *general validity*, which is a purely formal notion that makes no assumptions about the status of the descriptive names and predicates in the language, and so which conforms to the Quinean desideratum that logic should be ontologically neutral. Real-world validity is truth in all models *in which the descriptive predicates and terms are all admissible*. If the admissibility of an expression has contingent presuppositions, then the rule of necessitation will fail to preserve real-world validity. In particular, in Jacinto's strongly Millian logic, a name is admissible only if it has a referent, so sentences of the form $\exists x x = t$ are real-world valid, but an admissible name might refer to something that exists only contingently, in which case the necessitation of the existential claim will be false.

The decision we made in our version of quantified first-order modal logic (the decision to allow for primitive singular terms that were admissible but non-referring) avoided this particular consequence, but as noted in that earlier discussion, when we move beyond the first-order theory, the problem returns, since in the higher-order language, the admissibility of descriptive predicates will have ontological presuppositions that can be expressed, and that may be contingent. In particular, an instantiation of our general comprehension schema (e.g., the following instantiation of the principle X: $\forall x \exists f \Box \forall y(fy \leftrightarrow Rxy)$ (where 'R' is predicate constant) will be real-world valid, since it follows from the generally valid comprehension principle, together with the real-world valid existential claim, $\exists r \Box (\forall x)(\forall y) \Box (rxy \leftrightarrow Rxy)$, that is presupposed by the admissibility of 'R'.[19] More generally, any instance of the schema M*, without the restriction that ϕ be a pure sentence, will be real-world valid, since any primitive predicate will be admissible only if it corresponds to a property or relation, and this implies that all of the required existential premises will be true in the actual world. But as Jacinto observes, real-world validity is not closed under the rule of necessitation. So the necessitation of an instantiation of M* will not be either real-world valid or generally valid.

Timothy Williamson defends the general validity of the strong comprehension principles that result from the necessitation of the instances of the schema (M*), and this thesis immediately entails that everything that exists necessarily exists, and that nothing could exist other than what does. He argues that rejecting the necessitation of the (applied) comprehension principles "has a cost: it obstructs normal second-order reasoning in modal contexts." In particular, "the failures of comprehension undermine the free use of the kind of reasoning that provides the most powerful

[19] Aside from issues about the contingent presuppositions of the admissibility of names and predicates, one way to force a distinction between real-world and general validity is to introduce into the language an actuality operator, a sentential operator '@' with the following semantic rule: $[\![@\phi]\!]_w = [\![\phi]\!]_a$ for all w. With this operator one can construct sentences such as any instance of the schema $(\phi \rightarrow @\phi)$ that are true in all models because true in the actual world of all models, but not true in all possible worlds in all models. So they are real-world valid, but not generally valid. As Jacinto observes in his discussion of strongly Millian modal logic, the necessitation of the schema (C*) will not be valid in a language that contains an actuality operator.

reason for introducing higher-order logic in the first place."[20] An example Williamson uses to illustrate this point is reasoning about phase spaces in physics, which are represented as possible states of a physical system. It is true that when one is reasoning about physical systems, one is not reasoning just about the actual world, but about a range of physically possible worlds. It is a contingent hypothesis that the world is as physics tells us it is, and (according to the contingentist), it is a contingent hypothesis that the properties and relational structures in terms of which physicists describe the world exist. But the properties and relations that constitute phase spaces with a particular structure will exist in a range of possible worlds, including all that are compatible with the theories and laws that the physicists are presupposing. So the existential premises needed to instantiate the pure comprehension principles (the instances of the schema C* with the specific predicates of physical theory) will be true, not just in the actual world, but in all of the worlds that the physicists take to be physically possible. Modal reasoning about those structures will face no obstruction from the hypothesis that they are not *metaphysically* necessary entities. More generally, the use of comprehension principles in reasoning within the scope of various restricted modal operators (e.g., reasoning about what is required or permissible) will be unproblematic so long as the relevant individuals, properties, and relations can be assumed to exist in the relevant possible circumstances.

5.5. A first-order theory and a threat of paradox

Let me sum up what we have done so far in this chapter. In section 5.1, I floated the ontological hypothesis that there are properties and relations expressed by the admissible predicates of an acceptable

[20] Williamson 2016, 732–733.

theory, and this quickly led to a hierarchy of properties and relations. Sections 5.2 and 5.3 explored a theoretical framework—a higher-order language—for representing the structure of this hierarchy. I considered Quine's reservations about this kind of framework and the dangers of equivocation that it posed, but I argued (in agreement with Quine) that with care the equivocations could be avoided if we accepted the ontological commitments that the higher-order quantifiers seem to be making, and if we distinguished the compositional role of an expression from its semantic value. In section 5.4 we considered a range of principles, stated in this kind of language, about the relations between entities of the various types that our ontological hypothesis posits, and argued that the theory is compatible with the contingentist picture that is suited to Quinean naturalism. In this final section I will consider what our theory would look like if we tried to formulate it as a first-order theory of this hierarchy of entities.

As I have emphasized, the type hierarchy that we are discussing is an *ontological* categorization, specified independently of any language used to talk about the entities categorized. It begins with a basic ontology of individuals (which could be any domain of things), and the whole type hierarchy is specified in terms of that basic domain and is dependent on it. That is, the membership in the type-categories of the hierarchy is specified relative to the basic domain with which we begin, using the criteria that we use to determine the admissible predicates for describing the things in that domain. The hypothesis that led to the augmentation of this ontology was that there are properties and relations corresponding to the predicates that are admissible to our basic theory, and we can make sense of this further ontological commitment only if we have a range of predicates for talking about these newly recognized things, and so we extended our hypothesis to include properties and relations corresponding to the predicates (such as the relations of exemplification between individual and properties) with which we describe the properties of and relation between the attributes of individuals. If we have succeeded in specifying a coherent ontology, it should be possible to construct a consistent first-order theory with the things in that ontology as its domain. Or, to put the point the other way around, if our attempt to construct such a theory results in a paradox, this would seem to show that we have not succeeded in specifying a coherent ontology. We have already, in setting up our higher-order language in an informal meta-language, quantified over all the types, and this informal meta-language seems to presuppose that we can talk about all of the things of the different types. We said, for example, that for any types $t_1 \dots t_n$, there is a type $\langle t_1 \dots t_n \rangle$. We could, in principle, state some of the closure principles discussed above in a first-order language, and we did so informally, saying, for example, that "for every binary relation and individual, there is a monadic property that something has if and only if it stands in the binary relation to the individual." Here is a statement that generalizes this rule, and so involves quantification over types, and over things of different types.

> For every type t_1 and t_2, and for every type $\langle t_1, t_2 \rangle$ relation and type-t_1 entity, there is type $\langle t_2 \rangle$ property such that any type-t_2 entity HAS that property if and only if the relation holds between the type-t_1 entity and the type-t_2 entity.

A first-order language in which such claims are formalized will have predicates for stating the generalizations—predicates that we take to be admissible. Can we assume that these predicates will themselves correspond to properties and relations? We have to be careful since there is a threat of paradox in a first-order formulation of a theory with this kind of ontology, a threat that is more serious than the threat of paradox that was considered, and defused, in the last part of Chapter 4. I think paradox can be defused here, too, or at least explained, but spelling out the argument helps to bring to the surface some questions about the relation between predicates and the attributes they express, about the identity conditions for attributes, and more generally about the relation between the subject matter of a theory and the language in which the theory is stated—or, to echo the Quinean manifesto, the relation between the "limning of the most general traits of reality" and the "clearest overall pattern of canonical notation."

The paradox, which was presented to me by Nick Jones,[21] is a variation on Russell's paradox, for attributes rather than sets. It assumes the following five theses:

(i) The language in which we state our theory of individuals, propositions, properties, and relations of all the different types is a first-order language with quantifiers ranging over a single domain containing things of all the types.
(ii) Predicates that are definable in terms of admissible predicates, using the variable-binding machinery of the first-order language, are admissible.
(iii) For any predicate that is admissible in this language, there is a property or relation that it expresses, and that is in the domain of the theory.
(iv) There is an admissible binary predicate of *exemplification* saying, of entities α and β, that α HAS β.
(v) If 'F' is a monadic predicate that expresses the property β, then for any α, Fx will be satisfied by α if and only if α HAS β.

Here is the argument:

(1) By (iv) and (ii), the following is an admissible predicate: $\lambda x \sim x$HASx'.
(2) So by (iii) there is a property expressed by $\lambda x \sim x$HASx'. Call it **b**.

[21] I thank Jones for spelling out, in discussion, this paradox for a first-order formulation of a theory of propositions, properties, and relations. The argument I will sketch is his argument, as I understood it. Thanks also for his permission for me to present my exposition of his argument, and more generally for his very helpful comments on a draft of this chapter.

(3) Then by (v), ($\mathbf{b}\text{HAS}\,\mathbf{b} \leftrightarrow \lambda x \sim x\text{HAS}\,x(\mathbf{b})$).

(4) Given that **b** exists, we can infer the following by abstraction:

$$(\lambda x \sim x\text{HAS}x(\mathbf{b})) \leftrightarrow \sim \mathbf{b}\,\text{HAS}\,\mathbf{b})$$

(5) Then from (3) and (4), we can infer the contradiction, ($\mathbf{b}\,\text{HAS}\,\mathbf{b} \leftrightarrow \sim \mathbf{b}\,\text{HAS}\,\mathbf{b}$).

The argument is valid, so at least one of the five theses needs to be rejected, but the paradox will be defused only if we have an explanation for why the rejected thesis or theses are mistaken. Consider, for example, thesis (iii). Quine, at a time when he was, at least tentatively, sympathetic to the hypothesis that there are properties and relations, asked himself, "Just what properties and relations are there?" and he replied:

> Well, [a] common sense partial answer is that every condition we can formulate (everything of the form of a statement, but with a variable in place of one or more signs of entities) determines an *attribute* of just the entities fulfilling that condition, a *relation* of just the n-ads of entities fulfilling that condition. (Roughly, whatever we say about an object attributes an attribute to that object; and so on). This is the principle of abstraction (*comprehension*).

But then he goes on to say "But not so fast. Even principle of abstraction untenable! (Paradoxes) There must be some conditions with no attribute or relation corresponding. A modicum of nominalism is thrust upon us *despite* our having begun with ever so willing a Platonism."[22] Quine gives no details about the looming paradox, but his condensed comment makes clear that he would respond to the argument we are considering by rejecting thesis (iii). Of course,

[22] These remarks are taken from informal notes that Quine wrote to himself in 1941 for a projected book, to be called *Sign and Object*. They are quoted in Verhaegh 2018, 84.

in the end Quine rejected properties and relations altogether, but if we go along with the "ever so willing Platonism" that accepts the hypothesis that there are properties and relations, but rejects thesis (iii), we need to explain why there is no entity corresponding to some intelligible condition. We motivated the hypothesis we are developing with something like Quine's common-sense answer to his question, "Just what properties and relations are there?" We began with the intuitive idea that a predicate is admissible when the world provides the resources for determining what is said in applying that predicate to something in the domain, and then we took what is said in such a predication to be the ascription of the property expressed by the predicate to the individual. To use Agustin Rayo's jargon,[23] we might have said that for a predicate to be admissible *just is* for it to express a property. The HAS predicate—the binary predicate of exemplification—seems to be clear enough, given that our domain contains both entities of some kind and properties that those things might have. What we would like to have is some explanation for why this predicate, or some predicate definable in terms of it, is an exception to the thesis that admissible predicates express properties or relations, or alternatively, perhaps we should look for an explanation of exactly why it is not so clear as it at first appears to be what condition is imposed by this predicate.

Here is one way we might explain why thesis (iii) should be rejected: As we noted above, the hierarchy of propositions, properties, and relations is specified relative to an initial domain for a first-order theory. The criteria for the admissibility of predicates of this initial theory are criteria specifically for predicates used to express propositions about the elements of that particular domain, and so we should assume that the properties and relations expressed by those predicates get their identity conditions from that domain. That is, one should think of the predicates of the theory as ways of distinguishing between things in the given

[23] See the discussion in chapter 1 of Rayo 2013 on the "just is" statement.

domain.[24] The first-order domain we start with might be a collection that contained no propositions, properties, and relations, and that is what is usually presupposed. In this case, the entities of the higher types will all be newly recognized entities—things distinct from those in the initial domain. The same point applies as we go up the hierarchy. Once the whole augmented ontology is specified, we could formulate a new first-order theory with that ontology as its domain, and with its own criteria of admissibility for its predicates. These new predicates will provide ways of distinguishing between a wider range of entities, and since properties and relations get their identity conditions from the range of entities they distinguish between, we cannot assume that the predicates of our new theory express properties and relations that are members of the type-categories specified relative to the original first-order domain.

But since there is no limit on the kind of thing we can theorize about in a first-order language, there is no reason why the initial domain couldn't be a domain that already contained propositions, properties, and relations. Could there be a *comprehensive* first-order domain that included from the beginning all of the properties and relations that are expressed by the predicates that distinguish between the members of that same domain—so that no new entities are introduced as we go up the hierarchy?[25] The paradox suggests that the answer must be no. If we accept this conclusion, it is still thesis (iii) that is being rejected in the diagnosis of the paradox, since what this conclusion implies is that

[24] Though it is a little more complicated than this. The predicates are taken to distinguish, not just between the things in the actual extension of the domain, but also between things of the kind that determine the domain. So as we emphasized in discussion of the possibility of a daughter of Kripke or a son of Clinton, our admissible predicates are expected to distinguish between possible things of the appropriate kinds.

[25] This is not the same as the question whether there can be a single domain of absolutely everything, but those who believe in the intelligibility of absolute generality will presumably accept that the absolute domain is initially comprehensive in the sense defined. Our reasons for disclaiming a commitment to a single domain of absolutely everything were independent of the threat of paradox, though this threat does add support to the disclaimer.

there can't be a first-order theory meeting the condition that for every predicate of the theory, there is a property or relation *in the first-order domain* that is expressed by that predicate. We can still accept thesis (i) since once a hierarchy of entities is specified in terms of a given initial domain, we can have a first-order theory with all of those entities as its domain. It is worth looking at this kind of formulation since it brings to the surface some questions about the relation between the different subdomains (the things of different types). These questions may help to explain why we cannot have a first-order theory with a domain that is comprehensive in the sense defined.

The questions I have in mind are about the identity relations between the entities of different types. In particular, are the type-categories disjoint, or might they overlap? In the context of the higher-order language, one tends to take for granted that each entity has its own unique type, and this presupposition is supported by the hypothesis stated just above that the properties and relations of each type get their identity conditions from the particular domain of things they are distinguishing between. But the higher-order theory does not have the expressive resources to say that the type-categories are disjoint, at least not with the identity predicate, since it is assumed that this predicate expresses a relation of type $\langle e, e \rangle$, and it assumes that sentences of the form $a = b$ are well formed only with names or variables that denote things of type *e*. The entities within any given type-category will have identity conditions—that was a prerequisite for allowing them into our ontology—and for each type, an identity predicate (different from '=') for that type will be definable in the higher-order language. For example if 'f' is a variable of type $\langle\langle\rangle\rangle$ (ranging over monadic properties of propositions) and p and q are variables of type $\langle\rangle$ (propositional variables), then the open sentence '$\forall f(fp \leftrightarrow fq)$' will say that the proposition that is the value of p is identical to the proposition that is the value of q. We could then state the postulate of the basic theory of propositions that equivalent propositions are identical with the following sentence of the higher-order language:

$$\forall p \forall q((\Box(p \leftrightarrow q) \rightarrow \forall f(f(p) \leftrightarrow f(q)).$$

Analogous identity predicates expressing a relation of type $\langle t, t\rangle$ for each t can be defined, and analogous identity conditions can be stated for properties and relations of any one type, but I don't see how to define, in terms of the other resources of the theory, an identity predicate of type $\langle t_1, t_2\rangle$ where t_1 and t_2 are distinct types. Could we introduce *primitive* predicates of type $\langle t_1, t_2\rangle$ for any t_1 and t_2 that expressed a relation of this kind? Such predicates should be admissible in the higher-order language as long as we assume that the questions about cross-type identities have answers, and that seems to be implied by the assumption that we have specified a coherent ontology. Perhaps the Quinean should say that our pre-analytic conceptions of property and relation do not settle these questions, but that our regimented theory should answer them by sharpening the pre-analytic notions that we find ourselves with in a way that is best suited for describing and clarifying the relevant phenomena. Some of the questions seem, intuitively, to be easy: Julius Caesar is not a property, a relation, or a proposition, and no monadic property is also a binary relation. But it may be less clear what to say about more abstract logical relations, in particular about the identity relation itself. Isn't there just one relation of identity expressed in the higher-order language by the different typed predicates? It does seem intuitively natural to hold that one is saying the same thing about triangularity and trilaterality when one says that they are identical as one is saying about the first postmaster general and Benjamin Franklin when one says that they are identical. More generally, there are many structural parallels across types: Is there perhaps just one relation of monadic exemplification? Isn't one saying the same thing about Socrates and wisdom when one says that

Socrates HAS the property of wisdom as one is saying about the property of wisdom and the property of being a virtue when one says that wisdom HAS the property of being a virtue? But if there is just one property of monadic exemplification, then why does the predicate 'HAS' of our first-order theory not express this property, which is a property in the domain of that theory?[26]

We have emphasized throughout that questions of ontology and questions of the admissibility of predicates are linked, according to the Quinean picture: predicates are admissible as ways of distinguishing between things in a given ontology, and a commitment to a certain ontology is a commitment to things that are distinguished by the range of predicates that are admissible in a theory of those things. But the Quinean story also emphasized that the identity predicate had a special status, which is what justified the claim that identity theory is a part of logic. Identity is not a predicate that is linked to any particular ontological commitment, but rather a predicate whose admissibility in a given theory is a general constraint on the acceptability of the ontology of that theory. But *identity*, and predicates definable in terms of it such as the predicate of existence, are still among the ways of distinguishing between, and saying things about, the entities in a given ontology that one is prepared to endorse, and it is open to us to choose to individuate properties and relations uniformly in terms of the particular domains to which they are applied. The identity *predicates* in our higher-order theory, like all predicates, are linked to a particular type, and we can take the relativity to types to be a feature of the relations that those predicates express. That is, the identity relation

[26] It should be noted that even if the exemplification predicate in the first-order theory does express a property that is in the domain of that theory, the argument of the paradox does not go through, since thesis (iii) will still be false: the *defined* predicate '$\lambda x \sim x$ HAS x' cannot be assumed to express a property that is in the domain since it will not correspond to any predicate definable in the higher-order language. The variables of the first-order formulation range over the whole domain, while predicates definable in the higher-order formulation are defined with variables ranging over only the entities of a given type.

expressed by the predicate that says that x is the same type-t object as y is a relation that holds only of type-t objects. This choice is not forced on us, but we may judge that it gives us the "simplest clearest overall pattern of canonical notation," which of course is not to be distinguished from the judgment that it is at least one way of "limning of the most general traits of reality."[27]

[27] Given the judgment that the type-categories are disjoint (which we can state in a first-order theory with the given hierarchy as its ontology), we can conclude that there could be, in the higher-order language, predicates of every type $\langle t_1, t_2 \rangle$ that say that the t_1 entity is identical to the t_2 entity. That is, predicates with this interpretation would be admissible. These predicates won't be of much use when t_1 is different from t_2, since predications made with them would be uniformly false.

6
Possible Worlds and Possible Individuals

You might be wondering at this point: what ever happened to possible worlds? We talked, in Chapter 4, about possible-worlds *models* for the proposed modal logic for the regimented languages in which propositions are expressed and described, but we haven't mentioned them in the discussion of the ontologies presupposed in theories formulated in a language with that logic. We started with a theory that makes an ontological commitment to propositions, and then, in Chapter 5, proposed to extend it to a theory that makes a commitment to a hierarchy of properties and relations, entities that necessarily conform to a rich set of structural conditions that can be spelled out in a higher-order modal language. This ontology, Quine would have thought, is very extravagant in its commitments, but not so extravagant as to contain a commitment to merely possible things, of any type. It does, however, allow that there might have been things (of any of the types) other than those there are. We gave examples of properties that (arguably) no *actual* individual could have, but that nevertheless *could* be instantiated, and this is possible only if there might have been things other than those there are. We further suggested that the same point can be made about entities of higher types: just as there might be *individuals* other than those that in fact exist, so their might be propositions, properties, and relations other than those there in fact are. If Agustin had had a sister, there would have been a property, *being identical to that particular person*, and there would have been singular propositions about that person. But given that he did not have a sister, there is no

Propositions. Robert Stalnaker, Oxford University Press. © Oxford University Press 2023.
DOI: 10.1093/oso/9780197647035.003.0007

property meeting that condition, and no proposition saying, about a possible person, that she is Agustin's sister. These conclusions about the possibility of things that do not in fact exist depend on substantive theses about what there is, and these theses can be disputed on metaphysical grounds, but it does not seem that logic, or a general theory about the essential interrelationships between individuals, propositions, properties, and relations should rule these conclusions out.

The general picture is this: The world presents us with domains of entities that can be organized into a hierarchy of kinds of things (individuals, propositions, properties, and relations, including properties of properties, relations between individuals and propositions, etc.). As we have emphasized, by "individual" we just mean things in some domain of a first-order theory that we are prepared to endorse. In the interpretation of our higher-order language, the domain of individuals is the starting point; the other entities over which the higher-order quantifiers quantify are things that correspond to the predicates that are admissible in the first-order language with which we begin, as well as predicates that are admissible for describing those further things. If the facts had been different (which means, if some of the true propositions had lacked the property of truth), there would still have been such an ontological hierarchy (which means: any maximal consistent proposition entails that there are things of the various types), but the particular things in any of the various categories might have been different. Our theory does imply that all the facts about the different ways that things might have been are determined by the facts about what actually exists—about what the actual individuals, propositions, properties, and relations are. But merely possible entities are not among the things (of any of these kinds) that the world presents us with. So how do we talk about what might exist but does not? We do it by going up a level and finding an actual property (of the appropriate type) that is in fact uninstantiated, but that might have been instantiated. That is, we find an actual way of characterizing a thing of the appropriate kind that is such that there is no thing that is characterized in that way, but which is such that there might have been one. (There is no actual thing that is Agustin's possible sister, but there is an actual *property*, the property of being Agustin's sister, that has an empty extension, but that might have had a non-empty extension.) Among the characterizations of this kind are properties of propositions for which there is no proposition in its extension.

Those of us who talk about possible worlds, but who reject David Lewis's modal realism, often point out that in using the expression "possible world," we are not literally talking about *worlds*, or universes. Possible worlds are ways thing might have been. I have argued that we should think of them as something like properties of a certain kind that the world as a whole might have had; more specifically, they are propositions that determine a *complete* way that things might have been. Our theory of propositions does imply that there are *maximal* propositions, where a proposition is maximal if and only if it is consistent, and entails every proposition that is consistent with it. Can't the maximal propositions serve as possible worlds? Yes, for some purposes, and under some conditions, but our theory allows for propositions that are maximal in this sense, but that still are, in a sense, incomplete. Say that a proposition **W** is *complete* if and only if it is maximal, and also satisfies the following witness condition: For every existential proposition entailed by **W**, there is a corresponding singular proposition entailed by **W**. But (continuing to make the contentious metaphysical assumptions sketched above) consider any maximal consistent proposition **P** that entails the proposition that Agustin has a sister. **P** will entail that there exists a singular proposition that would be true if and only if a certain particular individual is Agustin's sister, but since there is in fact no such person, there will in fact be no singular proposition fitting this description, and so no proposition at all that fits this description, and that is entailed by **P**. So while **P** is maximal, it is not complete, and (according to the metaphysical assumptions we are making), the world does not present us with the resources to complete it.[1] There do, however, exist *properties* of propositions that we can use to characterize a *kind* of proposition that is both maximal and complete, and that entails that Agustin had a sister. So even if there is in fact no proposition of this kind, we can say that there might have been such a proposition; in fact, there definitely would have been a proposition meeting this condition if Agustin had had a sister.

The Kripke model theory, with its domain of possible worlds, each with its domain of the possible individuals that would exist if that world were realized, does not seem to be compatible with the metaphysical hypothesis we are considering, according to which there might have been individuals, properties, and propositions other than those that actually exist. The reason is that the possible worlds of a Kripke model will all be *complete* propositions since their domains will include particular merely possible individuals. Our contingentist metaphysical picture seems to imply that the world does not present us with the entities we need to justify the ontological commitments that a Kripke model seems to be making.

My response is to accept this judgment, but to claim that we can still use Kripke models (with their domains that contain merely possible individuals, and "worlds" that are merely possible propositions) to represent the facts, including modal facts, that are determined by the actual things of all the different types that are presented to us by the world and the complex relationships between them. The nonactual possibilia that are parts of the Kripkean structures will be artifacts of the model, rather than things being modeled. But the artifacts will be playing an essential role in the model theory, and so we need to explain and justify that role.

The issue concerns the relation between the part of reality that is the subject matter of a theory and the Kripke structures that are used to model the relation between the language of the theory and that reality. I will try to clarify this relation by developing a simple

[1] The *true* maximal proposition, of course, *is* complete in this sense.

example. It will be a toy story, with a minimal finite domain of actual things, and a very limited range of admissible predicates, and of actual propositions. But it will be an example that allows for the possibility of individuals, propositions, and properties that do not actually exist, and so will illustrate the way in which domains that contain merely possible individuals as members enter into the model theory, even though there are no merely possible things in the reality being modeled.

Here are some of the facts (described informally) about the portion of reality that our toy theory talks about: There are just two things, two coins, which are flipped, both landing tails (though they might have landed any of the other ways two coins might land). Furthermore, the coins might not have existed at all. Instead, there might have been a pair of dice that were tossed and landed any of the ways that a pair of dice might land (1-1, 1-2, . . . 6-6).[2] In developing the relevant part of the example, and the way it is modeled, we can ignore the coin for now; the interest in the story is in the merely possible dice, and the propositions about what might happen to them. The only admissible predicates in our theory, or at least the only ones we will be concerned with, are those describing the ways the coins or the dice land. So a sentence that says that the dice exist, and that they land a certain way (say one 6 and the other 5) will express a maximal proposition. The proposition will be maximal because the dice are indistinguishable with respect to their capacities, so there is nothing more to be said about them, in specifying a possible situation, other than how they land. The capacities of the dice are represented by the alternative possible situations that represent the alternative ways they might have landed.

Since the dice might have landed any of the ways that a pair of dice might land, there are, it seems, two different ways of landing

[2] This toy model is a variation of an example Kripke uses to make a different point. In Kripke's story, the dice are actual dice, and his point is about the identification of individuals across possible worlds.

that fit the description, "one landed 6 and the other 5"—two ways that a certain maximal proposition might have been true. Each of the two dice might have landed 6, and each might have landed 6 while the other landed 5. If we label the dice "A" and "B," we might say that in one of these possibilities, it is A that lands 6, and B that lands 5, and in the other it is the other way around. But what do we mean when we label one of the dice "A" and the other "B"? There are no dice—there is only the possibility of there being a pair of dice. Which of "them" was the one that we labeled "A"? There is no proposition to distinguish them. We can't say that it is the one that landed 6 in the possible situation we are considering, if by "possible situation" we mean a maximal proposition, since our ontology allows for only one possible situation, in this sense, in which one of the dice landed 6 and the other 5. Still, a correct account (an adequate regimentation) of our story needs to recognize the sense in which there are two different possibilities in which the dice landed 6 and 5. One way to make this point, informally, is to say that if the dice had landed one 5, the other 6, then there would have been a different possibility in which one landed 5 and the other 6, but the other way around. We can describe this possibility in a regimented modal language. If "S" is the predicate for "landed 6," and "F" is the predicate for "landed 5," the sentence is as follows: $\exists x \exists y (Sx \wedge Fy \wedge \Diamond(Sy \wedge Fx))$. The proposition expressed by this sentence does not distinguish the two dice, which cannot be distinguished since they have the same capacities.[3] The job of the variables in a regimented language with complex predicates and quantifiers is to keep track of cross-reference, but the word "cross-reference" is perhaps misleading, since the bound variables are not referring to anything. What they keep track of is the interrelationships between argument places in a complex predicate, or a complex general statement. In informal

[3] A story in which the dice had different capacities—for example, one in which each might have landed either 5 or 6, but only one of them might have landed 3, we could use that difference to distinguish them.

statements in which we use ad hoc labels to describe a complex possibility we are using the labels in the same way. (Suppose the dice existed, and that one landed 6—call it "A"—while the other landed 5—call it "B." Then it would have been possible that it be the other way around, with A landing 5 and B 6 instead.)

In defining a Kripke model structure for our little story, we introduce names to represent, on a more global level, the kind of cross-reference relations that variables are used to represent in the individual sentences of the language of the theory. Just as there is no difference between the representational roles of "x" and "y" in any quantified sentence (i.e., if we permute them, the resulting sentence would say the same thing), so a Kripke model structure that permuted the merely possible things would be a representationally equivalent model. But the situation with Kripke models with names for possible things is more complicated than this makes it sound, since the *actual* individuals (the members of the domain of the actual world of the model) have a different representational role from the merely possible individuals that belong to the domains of other possible worlds. The domain of the actual world in an intended model is a domain of things found in the world that we choose to talk about. Those in the domains of other possible worlds that are not among the individuals in the domain of the actual world are there to model the relationships between the possibilities of there being things of the appropriate kind other than those there are. I can't name the dice, since there are no dice to name, and there are no facts (according to our story) that would distinguish a 6-5 possibility from a 5-6 possibility. But the situation with the coins is different. The two coins have the same capacities (just like the merely possible dice), and we cannot distinguish them by how they landed, since they landed the same way. But the world still provides the resources to distinguish them, since they actually exist. For each of the coins, there exists the property of being identical to that coin. This is not a point about the capacity of a speaker in the actual world to distinguish or to name something. It might have been that no person was in a position to distinguish two different actual things. (The same point would apply to pairs of coins that exist in a galaxy far, far away, where there is no one around to refer to them.) The issue is about what properties and propositions exist, not about what we or anyone is in a position to say or to think.

Just as the actual coins and the counterfactual dice are playing different roles in the model we construct for our story, so the actual world of the model has a different status from the merely possible worlds. The actual "world" of the model is an actual proposition—the one true maximal proposition—while some of the counterfactual worlds (the dice worlds) represent only the possibility of a proposition—a *kind* of proposition that would exist if the dice had existed and been tossed. Each possible world of the model represents a possible actual world, or more precisely, the possibility that there be a proposition other than the true maximal proposition that is maximal, complete, and true.

In one way, the possible worlds in a Kripke model structure, with their domains of individuals, are on a par, and we can be precise about the way that they are on a par: As long as (1) our model is used to interpret only a first-order language, (2) we treat singular terms as we have, and (3) the logical resources of the language contain only the operators we have given it,[4] then the distinguished actual world of the model structure is no different from the other possible worlds in the sense that we could interchange the actual world with any of the other possible worlds and still have a structure that met all of the constraints that the model theory requires. The idea is that each possible world is, from this neutral perspective, a potential actual world, and to view a counterfactual world as a potential actual world is to view its domain as a domain of things that exist—things that *would* exist if that world were the actual one. But we can acknowledge this sense in which our model is ontologically neutral

[4] In particular, there is no actuality operator.

while at the same time recognizing that the entire theory is a view of modal reality from the perspective of the actual world—the only world there is, if we mean by "world" the universe that corresponds to the true maximal proposition. The distinctions between possibilities that are being modeled are only the distinctions that the actual world provides the resources to make, and this includes the resources for describing the ways in which the resources for distinguishing between possibilities might have been different. That is, we must use actually existing propositions to describe counterfactual situations according to which there exist propositions different from those there actually are. The point is that the entire construction of a model of modal reality takes place in the actual world, and it gives us a view of modal reality only from the perspective of the actual world. This point is an application of the more general point that Quine continually emphasized throughout his philosophical theorizing: "It is understandable . . . that the philosopher should seek a transcendental standpoint, outside the world that imprisons [the] natural scientist and mathematician. . . . However, there is no such cosmic exile."[5]

In the artificial models constructed to prove completeness discussed in Chapter 4, extra names were introduced to ensure that every individual had a name, and some of these names were also the objects named. In the basic construction, the objects in the domains of merely possible worlds, and their names, were treated no differently from the objects and names for the actual things, reflecting the sense in which possible worlds are on a par. These models will better reflect the distinctive role of the actual world, and the ontological commitments that are being modeled, if we consider a description of the whole model structure in the restricted language of the actual world, L_0, the language whose vocabulary includes only

[5] This quotation is taken from unpublished notes written by Quine in 1944 for a projected book to be called *Sign and Object*. They are quoted by Sander Verhaegh in an epigram to Verhaegh 2018.

names that refer in the actual world.[6] The descriptions of the merely possible worlds in this restricted language will be partial and will not distinguish possible worlds that have the same representational role, worlds that represent the same maximal proposition. We ignore the names of the other languages, and so the members of the other domains: they are not among the things in the domain of the language/theory for which we are providing a model.

The artificial construction for the completeness proof was a first-order model: the only ontological commitments we were considering at that point—the only ontological commitments expressible in the first-order language—were to the members of a domain of individuals. The theories that the Kripke models are designed to model, of course, have predicates, and we assumed that the predicates are admissible. As we, following Quine, have emphasized throughout, the hypothesis that a predicate is admissible may have substantive contingent presuppositions, but these presuppositions will not be directly reflected in the ontology or in the logic so long as we remain on the first-order level. In Chapter 5, when we floated the hypothesis that there are properties and relations corresponding to the admissible predicates, we opened the door to the possibility that properties and relations, like individuals, might exist only contingently, and that there might exist properties and relations other than those that actually exist. This expansion of the ontological commitment brings to the surface further consequences of the fact that our perspective on what is possible is a perspective from the actual

[6] This sublanguage, as well as analogous sublanguages for each of the possible worlds, was defined in section 4.3 of Chapter 4. For any of the names in the language L^+ that were added during the construction, the sublanguage L_n included only those that were in the domain of the world *n*. The intuitive idea was that the new names introduced should be thought of as like those singular terms of an actual regimented language that are introduced as de jure rigid designators for a particular identified individual. They are therefore, admissible only if they have a referent. The more liberal assumptions of our modal language allowed for the possibility of singular terms that were admissible but did not refer. But these terms would correspond to terms introduced into the language by equivalence to a description.

world, and that the resources available to characterize counterfactual possibilities are restricted to things (individuals, properties, and relations) that are available there. The predicates with which our theories describe both the actual world and the merely possible worlds can only be predicates that are admissible *in the actual world.* So while the artificial model we constructed for the completeness proof introduced some inadmissible names to the language, the *predicates* of that language were all presumed to be admissible in the actual world of the model. But just as we can use propositions about actual individuals to describe possibilities in which those individuals do not exist, so we can use the predicates admissible in the actual world to describe possibilities in which the properties corresponding to those predicates do not exist, and possibilities in which there exist properties that do not actually exist. Consider again the property of being Agustin's sister, an actual property that is uninstantiated, but that might have been instantiated. Now consider a possible situation in which Agustin himself did not exist—perhaps his parents had only daughters. In this possible situation, the property of being Agustin's sister would not only be uninstantiated—it would also be nonexistent. The property of *not* being Agustin's sister would also fail to exist in such a situation, but it *would* be instantiated there, since the property of not being Agustin's sister would apply to everyone in any world so unfortunate as to be without Agustin.[7]

The existence of the property of being Agustin's sister presupposes the existence of a particular individual—that is why it is a clear case (on the assumption that Agustin might not have existed) of a contingently existing property. But our logic should not assume that this is the only kind of property that could fail to exist. One might have to go further afield to find possible circumstances in

[7] Just to complicate the story, we might say that if Agustin's parents had had only daughters, then one of those daughters might have had an older brother, and that brother might have been Agustin. But we can describe that possibility only (in several steps) from the perspective of the actual world.

which paradigm purely qualitative properties (colors, for example, or properties studied and applied in fundamental physical theory) do not exist, but consider a possible world in which there was no electromagnetic radiation and also no conscious beings with phenomenal experience. Would color properties exist? One might be tempted to say that the property of being red would still exist—it would just be that nothing had it. We can agree that the property of being red would be uninstantiated in a world of this kind, but that is an answer to a question about the extension of an actual property in a counterfactual world (like the question about the extension of the property of being Agustin's sister in a world in which Agustin did not exist). For the property of being red to exist in the counterfactual world is for *that* world to have the resources to determine what a world would have to be like for that property to be instantiated, and it is not clear that a world of the kind I have described (no electromagnetic radiation, etc.) would have the resources to individuate the particular color properties that exist in our world. If a world fitting this description were realized, it might have the resources to provide a very general characterization of the structure of a world like ours, but still lack the resources to define the family of particular properties that have that structure.

The use of Kripke models to represent modal reality has great heuristic value. Judgments about possibilities are naturally described in terms of particular things, and worlds, that might exist, but don't, even though the very idea of an ontological commitment to things that don't exist cannot be taken literally. But the Kripke models are not of just heuristic value, since the merely possible individuals that are introduced in the characterization of the models are an essential part of a precise articulation of the theory of properties, relations, and propositions. Their role, as suggested above, is to keep track of the interrelationships between possible situations in which there are entities that do not in fact exist. This role shows itself in the details of the compositional semantics of the modal language, specifically the details of the semantics of variable binding, which requires reference to the merely possible individuals that (according to the contingentist metaphysics) are artifacts of the models. But the role of these artifacts is to describe the relational structure that is part of the reality being modeled, and we can generate a model of this kind using only the resources that are provided by this reality.

The modeling strategy that we use to interpret a modal framework is what might be called the supervaluation strategy. Supervaluations were first introduced by Bas van Fraassen in order to model failures of reference, but the notion was later applied more widely, most prominently to vagueness.[8] One starts with the judgment that there seems to be no fact of the matter whether certain statements are true or false, even though some complex statements with truth-valueless constituents seemed to *have* truth-values. To model such situations, the idea was to define a collection of arbitrary extensions of the truth-value assignments so that in each of the extended valuations all sentences get truth-values, and in which the classical compositional rules are satisfied.

In the simplest case of an application of the supervaluation strategy, the language being interpreted is an extensional propositional logic, and a classical model for the language is simply a truth-value assignment to the sentence letters, with the truth-values of the complex sentences determined by the truth-table rules. A partial model (an assignment of truth-values to some but not all of the sentence letters) determines a class of complete models, and a supervaluation model is a model that assigns true to all sentences that are true in all the complete models in the class, false if false in all, and neither true nor false if true in some and false in others. Classical sentential logic is validated, since a sentence is true in all supervaluations if and only if it is true in all classical valuations.

[8] See Van Fraassen 1966, and for the use of this strategy in an account of vagueness, see Fine 1975. For a later development of Fine's views about vagueness, see Fine 2020.

The strategy generalizes to richer languages that require models with additional primitive parameters (e.g., domains of individuals, possible worlds, temporal or causal structures). With any such case, one can define partial models that correspond to classes of complete models of the appropriate kind. One familiar case of this kind of modeling is a branching time model for a language with both tense and modal operators, where the modality is a time-dependent necessity, representing the facts that are settled, or determined at that time. In a determinate model of this kind, there would be a designated actual complete history through the tree, representing the actual path that history has and will take. But on a temporalist metaphysical view, the determinate facts, at any point, will be only the facts that are determined at that point, so the "actual world" is represented just by a designated node of the tree. The alternative branches from that point represent alternative possible futures, but (according to this metaphysical view) there is no fact of the matter about which way the future will develop. More generally, there would be no fact of the matter at any future, past or counterfactual point which way the future would develop from that point. A model of this kind will determine a class of histories, and we can define a fine-grained notion of "possible world" as a total path through the tree and a fine-grained notion of proposition as a class of complete histories. The temporalist model corresponds to a class of determinate models. The temporalist says that a proposition (in the fine-grained sense) is true at a given node if and only if it is true in all completions of a model where that node is the actual present, false if false in all such models, and neither true nor false otherwise.[9]

[9] See Thomason 1970 and MacFarlane 2003 and 2008. Thomason, with A. Gupta, later extended this kind of semantic theory to incorporate an account of counterfactuals in Thomason and Gupta 1980. I used this kind of theory to represent counterfactuals about objective chance in Stalnaker 2019. MacFarlane's general account of relative truth, developed in MacFarlane 2014, can be seen as an application of the generalized supervaluation strategy.

There are other examples. One that I have used as an analogy with the contingentist interpretation of modal model theory is a relativist interpretation of space.[10] In a simple Newtonian theory of absolute space, spatial relations can be reduced to intrinsic spatial properties: locations or spatial points. Relations are reducible in the sense that a specification of the absolute location of each thing in a collection of particles will determine all the spatial relations between those things. But according to the relativist, there are no locations, since there is no fact of the matter about whether the universe as a whole is in motion or at rest. The Galilean relativist observes that we can treat locations as artifacts whose role is to represent relational structure, using the Newtonian models, but then defining an equivalence relation between those models. The claim is then that the real properties and relations are those that are invariant for all equivalent models. So long as the equivalence relation is clear, the artifacts are adequately justified without an ontological commitment to entities in an intended model to which those artifacts correspond. Any case where a structure of relations is not grounded in the intrinsic properties of the relata is a case where this kind of modeling structure is appropriate. Some metaphysicians may follow Leibniz in accepting a very abstract metaphysical principle that all relations are reducible to intrinsic properties, and so may judge that we have not reached a fundamental level until will have found a way to ground all relations, but I don't see any a priori reason to accept such a principle.

Even if it is right that the supervaluation strategy is clear and coherent and does not make further metaphysical commitments, one might argue that Quinean methodology would justify taking on board the necessitist ontological commitments that the Kripke models seem, on a straightforward interpretation, to be making, at least once we have accepted the modal logic and the commitment to propositions. The contingentist agrees that the Kripkean model

[10] Stalnaker 2016.

theory is a fruitful theory for modeling these commitments, and if it is really true that "the quest of a simplest, clearest overall pattern of canonical notation is not to be distinguished from a quest of ultimate categories, a limning of the most general traits of reality," don't we have at least some reason to accept the metaphysics that comes with the simplest and most straightforward interpretation of that model theory? But as noted in Chapter 1, the Quinean manifesto is a rhetorical exaggeration, since Quine also makes clear that some choices about canonical notation are independent of substantive ontological and ideological commitments. The Quinean will agree that the acceptance of an overall framework/theory is an endorsement not only of the claims that that the theory makes but also of the questions that it raises. That is, a framework/theory may presuppose that certain questions have answers even if it does not say what those answers are. Consider the contrast between the Newtonian who is an absolutist about space and the Galilean relativist. The former theory raises questions about whether the universe as a whole is at rest or is moving at constant velocity through space; and, if so, in what direction and at what velocity. The judgment that these are bad questions that don't have answers is a reason (for one who believes that the Newtonian theory is otherwise an acceptable theory) to favor the Galilean interpretation. It need not be obvious whether a question has an answer, even if it is obvious that the theory does not at present provide a way to answer it. In some cases, it may be reasonable to judge that there are questions that have answers, even though we will never be able to know what they are (e.g., historical questions about the fine details of events that took place long ago). But the judgment (subject to correction, as always) that there are no answers to the questions that the Galilean relativist rejects should be intelligible even to the Newtonian who rejects it, and I think the same should be said for the contingentist who accepts the fruitfulness of the Kripkean models.

Among the questions raised by the necessitist metaphysics that the contingentist might judge to be bad questions (questions with no answers) are cardinality questions about merely possible things. We agree that Agustin might have had a sister, and the necessitist takes this to imply that there are some things that might have been Agustin's sister, but how many of them are there? The contingentist takes this to be a bad question, rejecting the presupposition that if Agustin might have had a sister, then there are things that might have been that sister. The contingentist who endorses the Kripkean modeling strategy will accept the intelligibility of questions about how many possible individuals we need in a Kripke model to represents a range of modal facts about Agustin. (For example, we need at least two to model the following regimented sentence: $\Diamond\exists x(Sax \wedge \Diamond\exists y(x \neq y \wedge Say))$, where 'S' is the *sister of* predicate, and 'a' names Agustin.) More generally, we can construct an infinite sequence of sentences with nested possibility operators and existential quantifiers that would require a model with infinitely many possible sisters, each in the domain of a different possible world. But how many do we need to model all the modal sentences we might construct that are true? That is a question about the resources of the language, and about whether the sentences that can be constructed from those resources are all intelligible.

Just as the Newtonian and the Galilean can reasonably disagree about whether the questions about absolute motion are good questions, so the contingentist and the necessitist can reasonably disagree about whether cardinality questions about possible individuals are good questions. Some of the arguments for necessitism that we have considered depend (I have argued) on equivocations, but there is a more purely metaphysical argument that has been given that cannot be criticized in this way.[11] This argument is based on the judgment that certain cardinality questions about collections of entities that might have been people or physical objects are good questions, and that the

[11] Fritz and Goodman 2017. The specific question they consider is how many incompossible things there are.

contingentist cannot answer them. I think this is an argument with some force, but whether it provides a reason for or against contingentism depends on what one thinks of the premise. A language may have too much expressive power if it allows one to say unintelligible things.[12]

[12] Possible physical objects and people are not mathematical entities, but there are cardinality questions about, for example, pure sets that I find puzzling. Is the question, "how many pure sets are there?" a good question? What about questions of the form, "are there at least α sets?" for some very large number α? Cf. Boolos 1998a.

References

Adams, R. 1981. "Actuality and Thisness." *Synthese* 49: 3–41.

Bacon, A. 2018. "The Broadest Necessity." *Journal of Philosophical Logic* 47: 733–783.

Boolos, G. 1975. "On Second Order Logic." *The Journal of Philosophy* 72: 509–527. Reprinted as Chapter 3 of Boolos 1998b.

Boolos, G. 1998a. "Must We Believe in Set Theory?' Chapter 8 of Boolos 1998b.

Boolos, G. 1998b. *Logic, Logic and Logic*. Cambridge, MA: Harvard University Press.

Carnap, R. 1950. "Empiricism, Semantics and Ontology." *Revue Internationale de Philosophie* 4: 205–221.

Dorr, C. 2016. "To Be F Is to Be G." *Philosophical Perspectives* 30: 39–134.

Evans, G. 1979. "Reference and Contingency." *The Monist* 62: 161–189.

Fine, K. 1975. "Vagueness, Truth and Logic." *Synthese* 30: 265–300.

Fine, K. 1977. "Properties, Propositions and Sets." *The Journal of Philosophical Logic* 6: 135–191.

Fine, K. 2020. *Vagueness: A Global Approach*. Oxford: Oxford University Press.

Florio, S., and N. Jones 2021. "Unrestricted Quantification and the Structure of Type Theory." *Philosophy and Phenomenological Research* 102: 44–64.

Fritz, P. 2016. "Propositional Contingentism." *Review of Symbolic Logic* 9: 123–142.

Fritz, P. 2020. "Being Somehow Without (Possibly) Being Something." Unpublished ms.

Fritz, P. 2021. "On Stalnaker's Simple Theory of Propositions." *Journal of Philosophical Logic* 50: 1–31.

Fritz, P., and J. Goodman. 2017. "Counting Incompossibles." *Mind* 126: 1063–1108.

Gallin, D. 1975 *Intentional and Higher-order Modal Logic*. Amsterdam: North Holland Press.

Goodman, J. 2016. "An Argument for Necessitism." *Philosophical Perspectives* 30: 160–182.

Goodman, N. 1983. *Fact, Fiction and Forecast*. Cambridge MA: Harvard University Press.

Jacinto, B. 2017. "Strongly Millian Second-Order Modal Logic." *Review of Symbolic Logic* 10: 397–454.

Jones, N. 2018. "Nominalism and Realism." *Nous* 52: 808–835.

Kripke, S. 1963. "Semantical Considerations on Modal Logic." *Acta Philosophica Fennica* 16: 83–94.

Kripke, S. 1979. "A Puzzle about Belief." In *Meaning and Use*, edited by A. Margalit, 239–283. Dordrecht: Reidel.

Kripke, S. 1980. *Naming and Necessity*. Cambridge, MA: Harvard University Press.

Kripke, S. 2013. *Reference and Existence: The John Locke Lectures*. Oxford: Oxford University Press.

Lewis, C. 1918. *A Survey of Symbolic Logic*. Berkeley: University of California Press.

MacFarlane, J. 2003. "Future Contingents and Relative Truth." *Philosophical Quarterly* 53: 321–336.

MacFarlane, J. 2008. "Truth in the Garden of Forking Paths." In *Relative Truth*, edited by M. Garcia-Carpentier and M. Kobel, 81-102. Oxford: Oxford University Press.

MacFarlane, J. 2014. *Assessment Sensitivity: Relative Truth and its Applications*. Oxford: Oxford University Press.

Prior, A. 1971. *Objects of Thought*. Oxford: Oxford University Press.

Quine, W. 1948. "On What There Is." *Review of Metaphysics* 2: 21–38. Reprinted in Quine 1961. (Page number citations from the latter.)

Quine, W. 1951a. "Two Dogmas of Empiricism." *Philosophical Review* 60: 20–43. Reprinted in Quine 1961. (Page number citations from the latter.)

Quine W. 1951b. "On Carnap's Views on Ontology." *Philosophical Studies* 2: 65–72. Reprinted in Quine 1966, 126–134. (Page number citations from the latter.)

Quine, W. 1953. "Three Grades of Modal Involvement." *Proceedings of the XIth International Congress of Philosophy*, 14. Amsterdam: North Holland. Reprinted in Quine 1966, 156–174. (Page number citations from the latter.)

Quine, W. 1960a. "Posits and Reality." S. Uyeda, ed., *Basis of the Contemporary Philosophy* 5, 391-400. Reprinted in Quine 1966, 246–254. (Page number citations from the latter.)

Quine, W. 1960b. *Word and Object*. Cambridge, MA: MIT Press.

Quine, W. 1961. *From a Logical Point of View*. 2nd ed. Cambridge, MA: Harvard University Press.

Quine, W. 1966. *Ways of Paradox and Other Essays*. New York: Random House.

Quine, W. 1975. "The Pragmatist's Place in Empiricism." In *Pragmatism: Its Sources and Prospects*, edited by R. Mulnaney and P. Zeltner, 21–39. Columbia: University of South Carolina Press.

Quine, W. 1986. *Philosophy of Logic*. 2nd ed. Cambridge, MA: Harvard University Press.

Rayo, A. 2013. *The Construction of Logical Space*. Oxford: Oxford University Press.

Rayo, A. 2021. "Beta Conversion and the Being Constraint." *The Proceedings of the Aristotelian Society* supplementary volume 95, 253-286.

Sider, T. 2011. *Writing the Book of the World*. Oxford: Oxford University Press.

Stalnaker, R. 1977. "Complex Predicates." *The Monist* 60: 327–339.

Stalnaker, R. 1994. "The Interaction of Modality with Quantification and Identity." In *Modality, Morality and Belief: Essays in Honor of Ruth Barcan Marcus*, edited by W. Sinnott-Armstrong, D. Raffman, and N. Asher, 12–28. Cambridge: Cambridge University Press. Reprinted in Stalnaker 2003.

Stalnaker, R. 2003. *Ways a World Might Be: Metaphysical and Anti-metaphysical Essays*. Oxford: Oxford University Press.

Stalnaker, R. 2008. *Our Knowledge of the Internal World*. Oxford: Oxford University Press.

Stalnaker, R. 2012. *Mere Possibilities: Foundations of Modal Semantics*. Princeton, NJ: Princeton University Press.

Stalnaker, R. 2016. "Models and Reality." *Canadian Journal of Philosophy* 46: 709–726.

Stalnaker, R. 2019. "Counterfactuals and Probability." Chapter 11 of R. Stalnaker, *Knowledge and Conditionals*, 182-202. Oxford: Oxford University Press.

Thomason, R. 1970. "Indeterminist Time and Truth-Value Gaps." *Theoria* 36: 65–90.

Thomason, R., and A. Gupta 1980. "A Theory of Conditionals in the Context of Branching Time." *Philosophical Review* 89: 65–90.

Van Fraassen, B. 1966 "Singular Terms, Truth Value Gaps and Free Logic." *Journal of Philosophy* 63: 481–495.

Verhaegh, S. 2018. *Working from Within: The Nature and Development of Quine's Naturalism*. Oxford: Oxford University Press.

Williams, B. 1978. *Descartes: The Project of Pure Inquiry*. New York: Routledge.

Williamson, T. 2003. "Everything." *Philosophical Perspectives* 17: 415–465.

Williamson, T. 2013 *Modal Logic as Metaphysics*. Oxford: Oxford University Press.

Williamson, T. 2016. "Reply to Stalnaker." *Canadian Journal of Philosophy* 46: 727–734.

Index

For the benefit of digital users, indexed terms that span two pages (e.g., 52–53) may, on occasion, appear on only one of those pages.